D1084979

Oil Companies in the International System

The Institute and its Research Committee are grateful for the comments and suggestions made by Jack Hartshorn, Robert O. Keohane and Dankwart A. Rustow, who were asked to review the manuscript of this book.

Oil Companies in the International System

LOUIS TURNER

The Royal Institute of International Affairs
George Allen & Unwin

First published in 1978

ISBN 0 04 382020 4 Hardback

The Royal Institute of International Affairs
Chatham House, 10 St. James's Square, London SW1Y 4LE

George Allen & Unwin (Publishers) Ltd
Ruskin House, 40 Museum Street, London, WC1A 1LU

British Library Cataloguing in Publication Data

Turner, Louis
Oil companies in the international system.
1. Petroleum industry and trade – Political
aspects 2. International business enterprises –
Political aspects
I. Title
II. Royal Institute of International Affairs
338.8 HD9560.5 77–30633

ISBN 0 04 382020 4

Printed in Great Britain
by Billing & Sons Limited,
Guildford, London and Worcester

Acknowledgements

During the four and a half years in which this book has evolved, I have become indebted to numerous people in a variety of ways. Some have granted me interviews. Others have typed or edited versions of the manuscript, commented on various drafts, or even, in one case, scoured the barrows of London on my behalf for rare and fascinating second-hand books on the oil industry. You will each know how you have contributed to this book, so thanks to: Morris Adelman, Jack Austin, Lord Balogh, Bill Barndt, Richard Barnet, Jim Bedore, Robert Belgrave, Joel Bell, Jean-Jacques Berreby, John Bird, Stephen Blank, Jean Boddewynne, Jacques Breil, Jack Bridges, Georges Brondel, James Buck, Denis Bulman, John Campbell, Monte Canfield, Guy de Carmoy, Marcello de Cecco, Bernard Cazes, Geoffrey Chandler, Joe Collins, Mel Conant, Philip Connelly, Ronald L. Danelian, Joel Darmstadter, Jerry Davis, J. F. Dean, J. G. Debanné, M. Desprairies, Bill Diebold, Paul Dimond, Andrew Ensor, Guy Erb, Jack Evans, Marilyn Evans, Judy Eversley, Edward Faridany, P. E. L. Fellowes, Nicholas Fenn, Ronald Ferrier, Mike Fliderbaum, Paul Frankel, Larry Franko, Alton Frye, Theodore Geiger, Richard Gilmour, George Glass, Walter Goldstein, William C. Gower, Alan Gregory, Margaret Greenhaugh, Don Guertin, Wolfgang Hager, Adrian Hamilton, Mary Hargreaves, Selig Harrison, Jack Hart, Jack Hartshorn, Jack Hays, John Henry, Richard Holland, David Housego, Robert Hunter, Edith Hutton, Bob Keohane, Kerryn King, Wilfrid Kohl, Ken Knubel, Ed Krapels, Bernard Kritzer, Hans Landsberg, Paul Laudicana, Remy Leveau, Sue Lewis, John Lichtblau, W. E. Lindenmuth, Al Litvak, Mike Lloyd, Robert Mabro, Bob McCoy, Thorold Masefield, Chris Maule, A. J. Milward, Edward Mitchell, Ted Moran, E. P. Mortimer, Noel Morton, Gill Mountain, Ronald Müller, Ken Myers, Paul Neville, Hugh Norton, Joe Nye, Peter Odell, Pella O'Neill, Christopher Patey, Jean Pell, Edith Penrose, Robert Perlman, Alan Peters, John W. Peters, Colin Phipps, David Powell, Jack Rafuse, Myer Rashish, R. G. Reid, Al de Russo, Dan Rustow, Anthony Sampson, Karl Sauvant, John Sawhill, Brian Segrue, Andrew Shonfield, Leonard Silk, Ian Skeet, Ian Smart, Dan Snook, Helena Stalson, Donald Stephenson, Bob Stobaugh, John Stokes, Susan Strange, John Surrey, P. G. Symington, Michael Tanzer, Karen Taylor, John Treet, Philip Trezise, Yoshi Tsurumi, Joy Turner, Teri Turner, Ray Vernon, Steve Warnecke, Nabiha Watson, Michael Watts, John West, Bill Whitman, John Wilhelm, Mira Wilkins, Mason Willrich, Joseph Yager.

My thanks also go to the Ford Foundation which funded the project on Transnational Relations of which this book is a product; also to Chatham House which has been a happy, stimulating place within which to work – with special thanks to the library and press library staff who have been invaluable allies, and Judy Gurney who has been a painstaking editor.

The brickbats, though, are all mine.

To Jean, with love

Contents

Glossary

AGIP Italian oil and gas company controlled by ENI (*q.v.*).
AIOC Anglo-Iranian Oil Company; *see* BP.
Aminoil American Independent Oil Company.
Amoco *See* Indiana Standard.
ANIC Italian refining and chemical company controlled by ENI (*q.v.*).
Anglo-Iranian *See* BP.
Anglo-Persian *See* BP.
AOC Arabian Oil Company; Japanese.
APOC *See* BP.
Aramco Arabian American Oil Company; originally a partnership of Socal, Exxon, Mobil and Texaco.
Arco Atlantic Richfield Company.
Atlantic Refining Company Merged to form Arco (*q.v.*).
BP British Petroleum; formerly known as Anglo-Persian (APOC) until June 1935 and as Anglo-Iranian (AIOC) until December 1954.
Caltex Company jointly owned by Socal and Texaco.
CASOC Californian Arabian Standard Oil Company; the forerunner of Aramco (*q.v.*); originally just a Socal subsidiary; Texaco joined as partner in 1936.
CFP Compagnie Française des Pétroles.
Chevron *See* Socal.
Conoco Continental Oil Company.
Creole Petroleum Corporation Exxon affiliate in Venezuela.
Deminex *See* Veba Group.
EEC European Economic Community.
Elf–ERAP Entreprise de Recherches et d'Activités Petrolières.
ENAP Empresa Nacional del Petroleo (Chile).
ENI Ente Nazionale Idrocarburi.
Exxon Offshoot of the original Standard Oil; was known as Standard Oil Company (New Jersey) until November 1972 (colloquially as Jersey or Jersey Standard; in some parts of the world affiliates are still known under Esso trade name.)
Gelsenberg *See* Veba Group.
IEA International Energy Agency.
IIAB International Industry Advisory Body (to the OECD Oil Committee).
Indiana Standard Standard Oil Company (Indiana); uses the Amoco name widely in its operations.
International Petroleum Cartel The cartel which the majors were alleged to have organised during the 1930s and 1940s. The term was given currency by the Federal Trade Commission investigations of the 1950s.
International Petroleum Commission A body proposed within the 1944 Anglo-American Oil Agreement; never activated.

International Petroleum Company Exxon subsidiary in Peru.

IPC Ambiguous – depending on the circumstances, can refer to International Petroleum Cartel, International Petroleum Commission, International Petroleum Company, or Iraq Petroleum Company; it most commonly refers to the latter.

Iraq Petroleum Company Consortium of majors active in the Middle East.

Japex *See* JPDC.

Jersey Standard *See* Exxon.

JPDC Japan Petroleum Exploration Company; sometimes known as Japex.

KOC Kuwait Oil Company. Joint venture between BP and Gulf.

MEEC Middle East Emergency Committee.

MITI Ministry of International Trade and Industry; Japanese.

Mobil Offshoot of the original Standard Oil; first known as Standard Oil Co. of New York (hence Socony); amalgamated with Vacuum Oil in 1931 to become Socony-Vacuum; changed name to Socony-Mobil in 1955 and to Mobil in 1966.

NIOC National Iranian Oil Company.

OAPEC Organisation of Arab Petroleum Exporting Countries.

OEEC Organisation for European Economic Co-operation. Forerunner of OECD.

OECD Organisation for Economic Co-operation and Development.

OPEC Organisation of the Petroleum Exporting Countries.

OPEG OEEC Petroleum Emergency Group.

Pemex Petroleos Mexicanos; Mexican state company.

Pertamina Indonesian state company.

Petrobras Petroleo Brasileiro. Brazilian state company.

Petrofina Belgian oil company (private).

Petromin Saudi Arabian governmental agency with responsibilities for oil.

PRC Petroleum Reserves Corporation.

Royal Dutch *See* Royal Dutch/Shell Group.

Royal Dutch/Shell Group Technically the correct way of referring to Shell; dates from 1907 when the Royal Dutch amalgamated its interests with the British company The 'Shell' Transport and Trading.

Shell *See* Royal Dutch/Shell Group.

Sinclair US oil company which merged with Arco in 1969.

Socal Standard Oil Company of California; uses the name Chevron widely in its operations.

Socony *See* Mobil.

Socony-Vacuum *See* Mobil.

Sohio Standard Oil Company (an Ohio Corporation). Under a 1970 agreement, BP is to take a majority shareholding.

Sonatrach Algerian state hydrocarbons company.

Standard Oil The original company dominated by John D. Rockefeller; dissolved in 1911 by Supreme Court decision; offshoots include Exxon, Mobil, Socal, Sohio, Indiana Standard.

Standard Oil (New Jersey) *See* Exxon.

Stanvac Joint venture between Exxon and Mobil.

TPC Turkish Petroleum Company; forerunner of the Iraq Petroleum Company.

Veba Group West Germany's leading group of oil companies; headed by Veba AG whose subsidiary Veba-Chemie concentrates on refining; Gelsenberg merged with Veba over 1974–5 and now concentrates on oil production; the group has a majority holding in Deminex, the exploration company.

YPF Yacimientos Petroliferos Fiscales; Argentinian state oil agency.

I
Introduction

This book takes a critical look at the myths which have built up around the relationship between national governments and giant oil companies. It looks askance at the arguments of industry defenders that the companies have just been good corporate citizens solely concerned with turning an honest buck, and at those of industry critics that the companies are some modern reincarnation of Attila the Hun, leaving a trail of toppled governments and impoverished economies in their wake. The truth, as usual, lies somewhere between the extreme views of the committed propagandists. The companies have undoubtedly wielded considerable economic power and have on occasion resorted to strong-arm tactics in their search for profits. Whether their behaviour is any 'worse' than that of the governments with which they have had to deal and, more particularly, whether they are significant actors in the international system, are questions which should concern those traditional students of international relations who tend tc ignore institutions which do not happen to be national governments on the grounds that only governmental activities significantly affect world politics.

The reason for taking a detailed look at the role of the giant, privately owned oil 'majors'[1] like Exxon and Shell[2] is simple. There are no larger examples of multinational companies, and of all non-governmental bodies, such companies seem to be the most powerful. Some are extremely large. Exxon, for instance, is now the world's largest industrial company in terms of sales, assets, net income and stockholders' equity. Its global sales reached $45 billion in 1976. Only in terms of the number of employees does it come down the list, with 137,000 employees comparing somewhat unimpressively with General Motors' 681,000. Supporting these giants are a large array of smaller private companies specialising in everything from ocean-going tankers to geophysical exploration. Encroaching on the majors' preserves are an increasing number of state oil companies, which are not yet as well known but are already impressively large in their own right. Sonatrach, Yacimientos Petroliferos Fiscales (YPF), Petrobras, Indian Oil, Pertamina, Pemex, Turkiye Petrolleri and the National

Iranian Oil Company (NIOC) are the largest companies in their respective economies (Algeria, Argentina, Brazil, India, Indonesia, Mexico, Turkey and Iran). In terms of sales NIOC was already the world's seventh largest industrial company in 1976.

The traditionalist in the study of international relations is unsure how to handle such companies. Those concerned primarily with strategic issues have tended to see a clear distinction between high politics involving issues like military security and global power balances, and low politics dealing with mundane issues like trade and tariffs. Even when turning their attention to the latter range of issues, they have rarely felt it necessary to analyse the role of non-governmental actors. Such views have been strongly criticised in recent years, as it became clear that a rigid distinction between high and low politics was becoming unrealistic. As Stanley Hoffman put it:

> The competition between states takes place on several chessboards in addition to the traditional military and diplomatic ones; for instance, the chessboard of world trade, world finance, of aid and technical assistance, of space research and exploration, of military technology, and the chessboard of what has been called 'informal penetration'. (1970, p. 401)

Aware that the issues calling for international diplomacy have been growing ever more complex, analysts have been increasingly concerned with non-governmental actors. In 1971, a seminal volume on transnational relations, edited by Joe Nye and Robert Keohane, looked at the international implications of a group of institutions and forces as varied as the Catholic Church, international migration, the Ford Foundation, revolutionary movements, international trade unions and multinational companies. They argued that the activities of such institutions form the heart of the transnational relations field, which they defined as the 'contacts, coalitions, and interactions across state boundaries that are not controlled by the central foreign policy organs of governments' (Nye and Keohane, 1973, p. xi).

By stressing the growth of such interactions, Nye and Keohane have undoubtedly helped steer foreign policy specialists away from a too-rigid conception of the central role of governmental bodies. However, there still remains a central ambiguity at the heart of their work. Certainly they have been able to show in their subsequent book, *Power and Interdependence*, that diplomatic transactions involving non-governmental actors have indeed been increasing rapidly in numbers, though, in the case of US relations with Canada and Australia, such transactions are still far from being the rule (Keohane and Nye, 1977).[3] What they are less successful in illuminating is the extent of the power such transnational actors have actually managed to exert. The fact that non-governmental bodies have commanded more diplomatic attention in recent years could merely indicate that in the past they were so much in control of the activities which most concerned them that governments either did not dare or need to get involved. This seems to be the reason why Gilpin, in the 1971 study edited by Nye and Keohane, claimed that the golden age of transnationalism was during the last century's *Pax Britannica* (Nye and Keohane, 1973, p. 56; Gilpin, 1976, p. 97), despite the fact that the two editors were arguing elsewhere that though such relations were not entirely new, 'their importance has been increasing in the years since the First World War' (Nye and Keohane, 1973, p. 398). Obviously, the two sides are talking about different things. Nye and Keohane are concerned with the frequency with which transnational interactions occur, while Gilpin is talking about the degree of freedom from governmental control which transnational actors may have had in the past.

The more one looks at the literature, the more one realises the lack of systematic studies of how the interaction of transnational actors with foreign policy establishments has evolved over the years. Should the transnational theorists be taken seriously because the scale of cross-frontier non-governmental interactions is increasing rapidly and hence deserves a more sophisticated treatment? Or have past analysts ignored the significance of non-governmental actors so that, by pushing the latter closer to centre stage, much of the history of the last fifty to a hundred years will have to be re-interpreted?

THE MULTINATIONAL DEBATE

In theory, the burgeoning literature on multinational companies should make it easy to decide whether such companies have been given too little attention in the past, but the more one reads, the less conclusive the facts appear concerning the extent of corporate influence on international relations. Certainly those closest to the transnational relations approach acknowledge that corporate power has to be taken into account. Kaiser talks of 'multinational corporations which often conduct their own foreign policy' (Nye and Keohane, 1973, p. 362). Nye and Keohane give prominence to Wolfer's belief that 'the Vatican [and] the Arabian-American Oil Company . . . are able on occasion to affect the course of international events' and imply that the international oil companies can affect the political stability of producing countries (1973, p. x).

There are critics of the multinationals who are more outspoken in their attacks on the alleged misuse of power by these companies. These critics have gained some support from the cases which have been given prominence in recent years, ranging from the ITT affair in Chile to the disclosures of massive political payments round the globe. It is easy to see why fears about the size and power of companies will not go away in the face of such revelations, and one is thus forced to take a serious look at claims such as Jay's that 'future students of the twentieth century will find the history of a firm like General Motors a great deal more important than the history of a nation like Switzerland' (1967), or at Caldwell's observations on the oil industry in East Asia:

> These magnitudes [GNP compared with company turnover] must constantly be borne in mind. If historians can deal seriously with the causes of wars in the past – when an empire might have fewer inhabitants and a thousandth of the wealth of a modern conglomerate – they ought not to dismiss without close consideration the role of these vast empires today. When we say 'Sukarno's measures increasingly restricted the operation of the American oil concerns', we are saying something much closer to 'the Kaiser's activities increasingly irritated the British in the running of their Empire' than we are to talking about a couple of grocers – or even grocery chains – having a price war. Standard Oil of New Jersey is much more powerful vis-à-vis Indonesia, however, than ever the British Empire was against the German. (1971, p. 6)[4]

There are critics who make even stronger claims, arguing that companies are not merely powerful, independent actors but that they take an integral part in the formulation of the domestic and foreign policies of their parent, imperialist governments. Kolko claims,

> No one can regard business as just another interest group in American life, but as the keystone of power which defines the essential pre-conditions and functions of the larger American social order, with its security and continuity as an institution being the political order's central goal in the post-Civil War historical experience. (1969, p. 9)

In the eyes of Kolko and his fellow revisionist historians, this corporate domination extends to the foreign policy sphere. They have tested their theories in extremely detailed analyses of the origins of the Cold War, arguing that this did not spring from Soviet expansion, but from the fact that US policy makers were determined to extend the areas of the world which would be safe for capitalism – a policy whose ancestry can be traced at least as far as the Open Door declaration of 1899 (Fleming, 1961; Williams, 1962; Alperovitz, 1965; Horowitz, 1965; Kolko, 1968, 1969; Maddox, 1973).

Kolko and the other revisionists base their arguments on detailed historical groundwork, albeit sometimes controversial. Most of the multinational debate takes place, however, in something approaching a historical vacuum, though the situation is slowly improving as authors such as Pearton (1971), Pinelo (1973), Wilkins (1974), Goodsell (1974), Moran (1974), Tugwell (1975), Anderson (1975) and Kent (1976) have begun to accumulate the kind of historical record which is necessary before reliable judgements can be made. However their work is only slowly starting to appear in the wider debate on corporate power and, though it is probably fair to say that most mainstream academics, including most of the transnational relations theorists (Keohane and Ooms, 1975, p. 183; Nye, 1975, pp. 125–34; Gilpin, 1976, p. 144) are now increasingly discounting an independent role for multinational companies, their arguments still tend to rest on the same inadequate historical records as those used by company critics. As a result, their rebuttals are far from definite. Vernon, for instance, acknowledges the justice of some radical claims (such as that the US marines were used to defend corporate interests) but makes a negative defence by pointing to cases where US corporate interests were distinct from those of the State Department:

> An unending succession of proceedings against large international enterprises since [the early 1950s] has contributed even more to confounding any simple hypothesis concerning business-government relations. Whatever the underlying rules of associations between US businesses and US government may have been – if indeed there were any – they evidently permitted frequent deviations from such simple rules as 'what's good for General Motors'. (1971, pp. 209–10)

Any analysis of the role of multinationals must obviously take Vernon's counter-examples into account, but it is disappointing that his implicit denial of any close relationship between companies and government ('if indeed there were any') does not rest on a firmer base than a justifiable riposte to radical oversimplification and a very brief analysis of US foreign economic policy, which he himself admits is not particularly illuminating in defining the scope of the corporate-government relationship (1971, pp. 210–14).

The most thorough champion of moderation is Kindleberger, but his defence is not properly documented and he makes a mistake made by many of us in the late 1960s and early 1970s of underestimating the extent to which some companies still use corruption or secret service links to further their ends. He does, however, do more than attack the radical arguments as too sweeping by looking at some of the factors which must be considered before making final judgements on the relative power of companies and governments. He argues that the effectiveness of government pressures increases proportionately to the size of the company assets at stake and to the degree to which such governments are company parents, that is to say they control the country from which a company originates and in which the bulk of its management is resident. He further suggests that the bargaining strength of companies relates to their relative positions of monopoly in addition to their hold on specific areas of technology (1969, pp. 152, 156, 193).

The present debate is thoroughly unsatisfactory, as the defenders of the multinationals seem to win their argument very much by default. There is a crying need for a wide-ranging study of at least one major industry designed to uncover the variety of interactions and patterns of influence between the relevant companies and the governments controlling the territories in which they function.

This book is an attempt to meet this need by taking a critical look at the wider political relationships of the oil majors. Now, obviously, oil is an exceptional case but, partly because of this, it is a relatively well-documented industry. As company apologists point out, much of the documentation has been done by detractors, but there are also some good solid historical volumes covering Jersey Standard and Shell, and a number of studies by people like Penrose, Jacoby, Longrigg, Odell, Hartshorn, Adelman, Tugendhat and Hamilton which provide far more analytical data about the political and economic activities of these companies than is found in connection with other industries.[5]

This documentation makes it possible to consider the extent to which foreign policy establishments have actually concerned themselves with the industry. How far have the traditional oil companies been para-governmental bodies, with the implicit function of providing their parent governments with supplies of secure, relatively cheap energy? To what extent are companies and governments monolothic entities, or mere amalgams of special interests, often pulling in different directions? In brief, what light can we throw on the cobweb of relationships involving the oil industry, be they between governments, companies, individuals, grass-roots pressure groups or intergovernmental agencies?

METHODOLOGY

Before beginning this study, two points need to be made about methodology. First, it would be very easy to rig the conclusions by only considering apparently 'reasonable' hypotheses – such as that the majors have possessed marginal influence, providing a source of pressure which has added to the richness of diplomatic history without being a dominant force. Although these are, in fact, the eventual conclusions of this book, they were reached only after dismissing much more extreme hypotheses about the power of the majors. Some readers will doubtless be offended that I have deigned to consider apparently sensationalist charges, such as that the majors may have backed opposite sides in Latin American wars or engineered political coups. The whole question of corporate power is an emotive one, and popular suspicions about the extent to which companies have manipulated governments run deep, particularly in the Third World. So conclusions must not only be reached, but must be seen to be reached in a painstaking fashion. There is no value in conclusions which can be immediately written off because certain hypotheses, however unlikely, have not been examined. What I have tried to do is to examine a wide range of hypotheses about the nature of corporate influence on world affairs, thus putting the onus on critics to disprove my conclusions.

The second point which must be made is that the historical record on which this book is based is still extremely patchy, reflecting the fact that oil historians have not been overly concerned with the interests of political scientists, while the fact that the latter have not often considered the role of the oil companies is the specific reason for this book. Given this historical vacuum, it is hardly surprising that a number of judgements made in this book are best described as educated guesses. Once again, some readers will be offended by this approach, but the fact remains that the fastest way to get historical lacunae filled is for some brash soul to stick his neck out by putting up a series of judgements which are worth challenging. This book represents such an attempt.

NOTES

1 The majors are the eight companies which by the 1940s had come to dominate the international oil trade – the companies we now know as Exxon, Shell, British Petroleum, Gulf, Texaco, Socal, Mobil and the Compagnie Française des Pétroles. If the latter company is excluded, they have also been known as the Seven Sisters.

2 Throughout this book companies are referred to by the names current at the time. Refer to the glossary to check the various changes in company names and for the meaning of abbreviations.

3 They show that transnational organisations are still only a factor in one-third of the cases under review. This counts incidents where the transnational organisation was lobbyist, target, catalyst, instrument or beneficiary of government actions.

4 In a review of one of my earlier books, Kindleberger has pointed out the fallacy of comparing figures for GNP (in which all double counting is eradicated) with company turnover figures (which are inflated by the value added by previous companies in the production chain).

5 The critical literature goes well back to the original muckraking era with books like that of Tarbell (1904). There was another upsurge of critical literature during the 1920s and 1930s.

2

Oil Majors and Parent Governments before 1939

For most of this century, producer governments have been of only marginal importance in the running of the international oil industry. It was the majors who controlled it, partly in their own right and partly with help from parent governments in a kind of implicit alliance of interests. It was this loose alliance which sustained the structure up to 1970, and within it one can identify fairly distinctive, alternating relationships. From the start of the serious search for oil, there was growing involvement of the parent governments, which was stimulated by the 1914–18 War and culminated in heavy postwar diplomatic in-fighting around oil rights, particularly in the Middle East. This initial period reached some form of conclusion with the 1928 Red Line Agreement[1] marking a truce between French, Anglo-Dutch and American interests striving to control oil development in the former Ottoman Empire.

The second period, from 1928 to 1939, is marked by relatively little parent government involvement, though there was some limited diplomatic activity stemming from the desire of one or two late-coming US companies to win stakes in Saudi Arabia and the various Gulf sheikhdoms. However, international oil issues were relatively low down the agendas of parent governments, and the three leading majors of the time, Jersey, Anglo-Persian and Shell, were sufficiently free of official supervision to establish the 1928 Achnacarry Agreement which, with various subsequent additions, regulated competition between the signatories.

The start of the 1939–45 War ushered in another era of extensive parent government involvement in the industry. The companies worked closely with the Anglo-American war effort but resisted plans for government intervention by the Roosevelt administration and an intergovernmental agreement between the USA and Britain to co-ordinate the world industry. A slightly more arm's-length form of active relationship between the companies and their parents sprang to life after the war in the Middle East, which reached its peak in the three years after Iran's nationalisation of Anglo-Iranian in 1951. By the time the dust of this episode had started to settle in 1954, the industrial structure of Middle Eastern oil had been re-formed one more time with sufficient stability to remain unchanged for the following decade and a half.

The period of 1954–70 was once again one of relatively low parent government involvement. Those who like to identify a 'Golden Age' for oil companies apparently independent of all governmental interference will normally point to the years on either side of 1957. They had come through the supply crisis of the 1956 Suez War with flying colours. Their apparent defeat of Iran's attempts to usurp Anglo-Iranian's role impressed most host governments with a need for caution. Their control over all the major sources of crude outside the Soviet Union was such that they could balance the expectations of one host government against another. Finally, oil was just starting to move into its current place as the dominant form of energy in the world, deposing coal from the throne it had held for so long. Admittedly, the formation of OPEC in 1960 was a sign of things to come, but, during the 1960s, the companies successfully resisted most of that body's initiatives. Parent governments could almost forget about the oil industry which, with a few isolated exceptions, was obviously being successfully controlled by the majors.

There is, then, no simple relationship between majors and their parent governments. The image of the free-ranging, politically independent oil company certainly holds true for some decades, but not for the beginning of the century, the 1940s or the 1970s. To find a pattern, one has to look at the ebbs and flows of the underlying power relationships of the various governments concerned. In times of stress, they have all appeared to see their national interest best served by their own oil companies having the widest range of operations. To some extent, the companies have been free to search for oil with or without the encouragement of their parent governments but this freedom has been constrained by the interests of key governments. In the early decades, the crucial decision makers were the great European powers, with their empires and relatively well-defined spheres of influence. Subsequently, the new American superpower was of growing importance in determining the fate of the companies. More recently it has become clear that the decolonisation process and the linked rise in Third World self-confidence has produced a new set of decision makers who can affect the oil industry.

If, then, we look at the five relatively distinctive periods in the history of international oil, a convincing picture seems to emerge. These five alternating periods relate to major shifts in the world's power balance: two of these shifts are the result of world wars while the third, in the 1970s, reflects a changed relationship between Europeans and North Americans on the one hand, and nations of the Third World on the other. The structure of the international oil industry altered as well. The first two power-shifts were marked by attempts of war winners to establish oil companies in the most promising areas of the world, over the objections of war losers. The last power-shift has been marked by the partial expulsion from production areas of the traditional companies and by a major redistribution of the wealth generated by the oil industry. In times of change, governments have been actively involved in staking claims for their companies and in fighting off claims by others. During the quiescent periods of consolidation of the 1930s and 1960s, once governments had redefined the power structure of the industry, they gradually withdrew to leave the field to their chosen agents, in the beginning the majors, and later, an increasingly mixed bag.

Although it is tempting to regard the oil companies as unfettered transnational actors, in fact they have won and held on to a good many of their most important concessions with the active backing and protection of parent governments. Seven Anglo-American-Dutch majors have dominated the Western world for most of this century partly because the British were the dominant power at the beginning of it, and the Americans for much of the rest. If the Japanese or Germans had won the last world war, the companies dominating the oil industry today would probably have been notably different. The fact that contemporary oil-producing governments have successfully challenged the hegemony of a handful of world powers suggests that the diversity of companies involved in the industry is going to grow rather than decline and that a multi-polar world will produce a heterogeneous oil industry. If the rest of the 1970s and 1980s are like the 1930s and 1960s, we should have a period of consolidation in which the governments now heavily involved will gradually withdraw, leaving various national champions, both private and state companies, to produce, process and market oil according to the deals which they themselves make.

PRE-1914

In the later decades of the nineteenth century (the American industry was in full swing in the 1860s, and the Russian one by the late 1870s) the imperial powers were not overly concerned with oil. It was not seen as a particularly strategic asset and had not been found widely – though the fortunes made in Pennsylvania and the Caucasus meant that there was no shortage of entrepreneurs involved. However, the colourful, cut-throat world of Rockefeller's Standard Oil, Marcus Samuel's Shell, Henri Deterding's Royal Dutch and the Nobel and

Rothschild interests in the Caucasus attracted more public interest when the British sensed the enormous contribution oil could make to naval power, and as European Great Power rivalry grew over the tempting oil rights to be prised from the Ottoman Empire.

The Royal Navy first experimented with fuel oil in 1899 (Henriques, 1960, p. 277) and by 1904 the First Sea Lord, Admiral Fisher, was convinced of its potential and determined that oil must be found promptly for the British (Tugendhat and Hamilton, 1975, p. 65). The government adopted a supportive stance toward William Knox D'Arcy's ventures in Persia, helping him find financial backing through the established Burmah Oil Company. With oil found and a European war looming, the British government became part owner of what was then the Anglo-Persian Oil Company (APOC). Convinced by enthusiasts like Winston Churchill that oil-fired battleships would be superior to those dependent on coal, Britain felt the need for a 'national champion' (Vernon, 1974, p. 11). The government stake in APOC was seen to provide financial advantages and security against foreign domination.

Oil concessions in the crumbling Ottoman Empire were diplomatically more complex. Here British, American, Dutch, German and Gulbenkian[2] interests were in competition. The US State Department supported the claim of an American citizen, Rear-Admiral Chester. The British government supported APOC's insistence that it be included in any deal which Gulbenkian, Shell and the Deutsche Bank put together. By 1914, the German government had accepted that a general reconciliation of European claims must be made to prevent the Turks from playing them off against each other. The result was the 'Foreign Office Agreement' uniting the European interests into the Turkish Petroleum Company – a clear case of how intertwined corporate and governmental interests had become, and of how British dominance could then ensure that aggressive diplomacy on behalf of its national champion APOC could wrest a dominant position of 50 per cent in such a potentially lucrative consortium (Church Report, 1975, pt 8, p. 496).

It would be misleading to suggest that this kind of active oil diplomacy was common. The USA had a thriving domestic industry, and was not particularly active in support of its nationals abroad. If anything, its pre-1914 official stance was hostile, particularly to the Rockefeller interests fragmented by the 1911 judgement concerning Standard Oil. Even within its Latin American sphere of interest, the USA was not sufficiently interested to stop an enterprising European like Weetman Pearson from establishing himself as an important oil producer in Mexico. Activism of the British government was more pronounced in this era.

1914–28

By the end of the 1914–18 War, no politician or strategist needed reminding that modern warfare relied heavily on oil. Public figures as distinguished as Clemenceau, Foch, Ludendorff and Curzon all went on record with stirring statements, the latter declaring that the allies 'floated to victory on a wave of oil' (Sampson, 1975, p. 60). Strategists had amended their war plans to include schemes for the seizure of available oil sources. Richard O'Connor suggests that the Germans invaded Romania in order to seize the Ploesti field, and claims that the role of Caucasian oil at Baku may have been a factor in Ludendorff's decision to send Lenin back to Russia (1972, p. 218). Even if this is a trifle far-fetched, Ludendorff was definitely concerned about oil shortages, and his Turkish allies made one ill-fated attempt to capture Baku. The British put a disproportionate effort into various Middle Eastern campaigns finally, without much embarrassment, capturing Mosul (then seen as having oil potential) some days after the armistice with Turkey was signed (Trumpener, 1968, pp. 344–51; O'Connor, 1972, pp. 212–18; Tugendhat and Hamilton, 1975, pp. 71–2; Kent, 1976, p. 241).

Even before the end of the war, the victors began to mould a basic structure for the international industry which, with modifications in the late 1940s and early 1950s, was maintained for over forty years. At first, it was primarily a question of Britain and France finding ways to carve up oil rights of the former Ottoman Empire. Britain felt vindicated in its prewar concern with oil matters, and France was traumatically shaken out of a peacetime complacency about overseas investment which left the country utterly dependent on foreign oil companies for supplies during the war, as exemplified by Clemenceau's December 1917 appeal to Wilson: '[The allies] must not let France lack the petrol which is as necessary as blood in the battles of tomorrow' (O'Connor, 1972, p. 217). The humiliation of dependence left a lasting scar on the French psyche and the Compagnie Française des Pétroles (CFP) was created as a mixed public and private national champion to represent French interests in Iraqi oil.

This European carve-up of Middle Eastern oil interests (the Netherlands were involved through Shell and by the fact that Gulbenkian was their wartime petroleum adviser; Church Report, 1974, pt 7, p. 163) took place initially without the USA, and was very much an intergovernmental affair, with various corporate interests on the sidelines watching the commercial implications. In 1918, the French and British swapped French claims on the Ruhr with British claims in the Middle East. By the time of the San Remo Treaty of 1920, they agreed that Britain would have a mandate over

Mesopotamia, and that French interests would replace the 1914 German stake in the Turkish Petroleum Company (Church Report, 1975, pt 8, pp. 496–7).

With its strong domestic oil industry, the USA had not worried overmuch about the international scene. But in 1920 it went through one of the now recurrent oil scares that the world is running out of oil.[3] With the mass acceptance of the automobile and the extraordinary demand stimulated by the war, domestic US prices for crude oil shot up. The US Geological Survey made a number of statements suggesting that the country's oil position was precarious and a US Senate committee suggested that American companies were being systematically kept out of most of the more promising foreign concessions areas by the British government. Informed public opinion was particularly alarmed by a claim of a British banker, Sir Edward Mackay Edgar, that: 'the British position is impregnable. All the known oil fields, all the likely or probable oil fields, outside of the United States itself, are in British hands or under British management or control, or financed by British capital' (Denny, 1929, p. 18). The San Remo Treaty rammed the point home. American companies which were unable to get into Burma, India or Persia were now to be kept out of the most interesting areas in the old Ottoman Empire (Church Report, 1974, pt 7, p. 163; Tugendhat and Hamilton, 1975, p. 746).

The result was a diplomatic offensive, aimed particularly at the restrictive practices of the British. The State Department sent instructions to all US diplomatic and consular offices, reminding them of their duty to help the crucial search for adequate supplies of oil to meet future American needs (Denny, 1929, p. 19). More specifically, the Department sent a series of forceful notes to the British Foreign Office reminding it of the 'Open Door' policy (first devised to protect US commercial interests in China), arguing that all countries should have access to the potential oil lands of the former Ottoman Empire. They argued that the USA and US companies had played a major part in keeping the allied war machine supplied with oil during the war (which was true), that the legality of the claim of the Turkish Petroleum Company (TPC) to Mesopotamian oil was dubious (Gulbenkian concurred), and threw in an implied threat that Jersey might cut off supplies to Britain. In the face of such pressure the British, French and Dutch gave in and allowed American companies to join TPC (Church Report, 1974, pt 7, p. 163), though an acceptable formula was not agreed until 1928 and a further three years were required before the exact American membership of what became the Iraq Petroleum Company (IPC) was stabilised (Jersey and Socony remained, Gulf, Texaco, Sinclair, Atlantic, and Standard Indiana dropped out) (Church Report, 1975, pt. 8, pp. 498–9). The five companies which were involved in IPC, if we ignore the Gulbenkian interest, were 'majors' – APOC, Shell, Jersey, CFP and Socony – which reached dominant positions in the Middle East through the diplomatic efforts of their various parent governments. The fact that the American companies were privately owned reflected the private enterprise ethos of the Harding, Coolidge and Hoover regimes, though Senator Phelan of California in 1920 had called for the creation of a state-owned company along the lines of Anglo-Persian (Tugendhat and Hamilton, 1975, p. 76). However, since the USA controlled over 80 per cent of the world's production, taking Mexico into account, it was the only country with self-sufficient existing companies. In the case of Britain and France, with the exception of Shell, government intervention had been necessary to produce companies which were capable of competing with American firms.

National styles were thus emerging. The British, concerned with oil, used their global power to back their interests. The French were equally concerned but tended to come out less satisfactorily from oil diplomacy, thanks to their lesser world status. They were insistent that a purely French company must look after their interests, rather than an international firm such as Shell. The latter had always been a company with strong established markets and inadequate supplies of crude oil and could thus only develop new markets by signing contracts with existing competitive producers (and thus, incidentally, forcing the company to do its calculations more closely). The Americans started to assert their right to a major world role by backing the descendants of the old Standard Oil as well as newer competitors. The American drive on British oil preserves was related to the rivalry of these two countries in naval affairs which, according to a historian of Anglo-American relations, was one of the two questions of major importance in this relationship between the two world wars (Allen, 1954, p. 733). The US Navy saw British dominance over world reserves as a reason for reaching equality with the Royal Navy and thus viewed oil as something which might conceivably be worth fighting the British for. Eventually these concerns died away, to be replaced by the threat that an Anglo-Japanese alliance might leave the US fleet to fight two significant naval powers simultaneously.[4]

The period from the end of the First World War until the Depression was one in which parent governments played a key role in allocating the world's potential oilfields to different national companies. It was also one in which some of these governments took conscious decisions on the relationship between national interests and oil company interests which ranged from the *laissez-faire* approaches of the USA and the Netherlands (though the latter relied primarily on a single company, Shell) to the more interventionist approaches of France and Britain. Given the relative passivity of governing authorities in oil-producing areas, there was little scope for parent governments to play a role of giving defensive support to companies under commercial or political attack. The one case where this was called for in a major way was Mexico, where attack by post-revolutionary regimes on the legal basis of established concessions kept the State Department and the British Foreign Office actively involved (Wilkins, 1974). Lesser, but related changes came from other Latin American countries like Peru, which also called for some defensive American diplomacy (Pinelo, 1973; Goodsell, 1974).

In this establishing era, the companies raised oil issues to the diplomatic level, as in the case when US companies complained to the State Department about British and Dutch discrimination. It was necessary for parent governments to agree before the companies could enter certain territories or make arrangements amongst themselves. The negotiations over American entry into Iraq symbolise this period. The State Department led the way with formal representations to the European powers, leaving the companies, led by Jersey, free to work out an arrangement with Anglo-Persian and its partners in the TPC within the overall framework of the 'Open Door' policy (a Door Open to US interests, but closed to any later would-be entrants) (Church Report, 1975, pp. 33–5).

The influence of oil company interests on diplomatic activity between the Soviet Union and the West in this era is a subject of transnational relations which has hardly been explored. Both Shell (from prewar days) and Jersey (from 1920) had an interest in the oil industry which the Bolsheviks had expropriated. During most of the 1920s, these two companies, with a number of others, haggled with the Russians trying to get both supplies of cheap oil and compensation for their expropriated property. Depending very much on personalities, industry double-crossing and the vagaries of the state of negotiations, company policies veered between being willing to do business with the Soviet authorities to outright hostility. As the leading oil companies had a distinctive, powerfully expressed (if inconstant) set of goals it should be possible to determine the impact of the oil companies on the policies of their parent governments towards the Soviet Union. Certainly Shell and Jersey had distinctive private 'foreign policies', Jersey because it bought the claims to Nobel's pre-revolution interests in 1920, and there is an allegation (without any supporting evidence) that Sir Henri Deterding was involved in a curious incident (the Arcos Raid of 1927) which led to the British breaking off diplomatic relations with the Soviets (Denny, 1930, pp. 291–2; Sampson, 1975, pp. 69–71). If the oil companies influenced the foreign policies of parent governments, there should be evidence of such influence here. I suspect that a re-interpretation of this period would upgrade the importance of oil company pressure on American and British policy makers at the time.

PARENT WITHDRAWAL: ACHNACARRY AND THE GREAT DEPRESSION

From the high point of the early 1920s when the State Department was insisting that the Open Door principle be applied to Iraq, the US government slowly lost interest in the international oil industry, reacting increasingly rarely to company cries for help. This was partly because new oil discoveries in America, such as the great east Texan discovery of October 1930, turned the supply of oil into a glut at the very time that the world was moving into the Great Depression. There were fewer calls for help after the Anglo-French-American agreement on IPC, as the declining number of companies interested in non-American production were satisfied with existing arrangements. Furthermore, with the exception of Bahrain and Kuwait, Western governments were not concerned with the Arabian Peninsula, so that the oil companies were left to wheel and deal with each other and with local authorities in relative freedom. This relaxed atmosphere was evident in Socal's entry into Saudi Arabia, which was a relatively straightforward case of competitive negotiation against IPC for the approval of King Abdul-Aziz. The British and American governments were nowhere to be seen. In fact, it was not until after the oilfields were officially opened in 1939 that the first US diplomat was accredited to Saudi Arabia – and he was the US minister to Egypt. As far as US officials were concerned during most of the 1930s, the oil industry abroad was of little interest. Of far greater importance was the need to help the domestic industry survive a fall in price from $1·30 to 10 cents a barrel. The most important initiatives were controls on oil production by state governments. This was the era in which the Texas Railroad Commission came to regulate the heart of domestic US production, backed up by the 1935 Connally 'Hot Oil' Act. Foreign oil was a threat and its importation was controlled (Sampson, 1975, pp. 74–7; Tugendhat and Hamilton, 1975, pp. 93–4).

The non-communist world was thus effectively cut into two separate markets, the more-than-self-contained American one, and the outside-United-States-one, in which a handful of companies struggled to control price and overproduction by co-operation among themselves. This was an age of international cartels[5] and the fact that Jersey, Anglo-Persian and Shell masterminded the Achnacarry or 'As-is' Agreement of 1928 was not unusual by contemporary industry standards. However, this clandestine agreement and its successors produced an anti-competitive ethos reflected in joint ventures like Caltex, Stanvac, Shell-BP in Nigeria and the Shell-Exxon exploration venture in Europe, which lingered into the 1960s and 1970s (Church Report, 1974, *IPC*, pp. 7–16; Tugendhat and Hamilton, 1975, pp. 97–111).

The decade after Achnacarry was probably the period in which the international oil industry was freest from any form of government control although market conditions were not as buoyant as those in the late 1950s and early 1960s (hence the latter is more generally judged to be the industry's 'Golden Age'). These companies were, for sure, somewhat nervous of the US antitrust authorities and based the administration of the cartel in London, but to all intents and purposes, they were free to manipulate world production and markets as they saw fit. For instance, if Indiana Standard chose to sell its Venezuelan offshoot, Creole, to Jersey, no government was going to say nay. Nor did anybody object to the entry of Jersey, Texaco and Socal into the Far East via the Stanvac and Caltex joint ventures (Church Report, 1974, *IPC*, pp. 129–30).

However, US companies in particular still had occasion to call on their parent government for help especially if they were outside the charmed circle of the IPC and its restrictive Red Line Agreement. There were still British spheres of influence such as Bahrain, which fell within the Red Line and was thus only open to a newcomer like Socal (IPC was not willing to let Gulf develop Bahrain and still remain a member of the Iraqi consortium). The British government tried to keep Socal out, but the State Department intervened to achieve a compromise whereby Socal entered Bahrain through a Canadian subsidiary. This formula was not unknown elsewhere: Jersey was involved in Peru through a Canadian company, the International Petroleum Company, which originated as a means of buying the British company which owned the relevant concessions (Pinelo, 1973, pp. 10–14; Sampson, 1975, p. 89). Anglo-American diplomacy also took place over Gulf's entry into Kuwait, when the British government's position was complex, sometimes helping Sheikh Ahmad to get better terms out of Anglo-Persian, but more often trying to stop the incursion of American interests by methods which included at least one rather dubious statement about the alleged insistence of the sheikh on a British control clause in any contract. Finally, after yet another Open Door appeal by the US State Department, Gulf and Anglo-Persian came to terms with the Kuwaitis (Chisholm, 1975, pp. 17, 21; Sampson, 1975, p. 92).

However, as the shape of the industry started to settle down, the problem for parent governments of protecting companies from turbulent host governments raised its head. For the British, this tended to be a case of backing British companies whatever the merit of the dispute. Earlier in the century they had provided troops to protect Anglo-Persian's workforce from local tribesmen (Elwell-Sutton, 1955, p. 21). Similarly, in 1924, they had dispatched a warship to protect a British oil refinery in Mexico (Cable, 1971, p. 182). Two cases in which government protection was offered to British companies during the 1930s were the disputes between Shah Reza and Anglo-Persian over the re-negotiation of the company's concession terms, and that between Shell and the Mexican government which expropriated both Shell and Jersey in 1938. In the former case, the navy dispatched a gunboat, without too much effect on the shah's composure. In the latter, the British broke off diplomatic relations as a protest, again with notably little effect – though the Second World War intervened in this particular dispute (Elwell-Sutton, 1955; Turner, 1973, pp. 77–80).

The State Department was more ambivalent in its attitudes to US companies. Certainly in the early 1930s there was not much hesitation about dusting off the Open Door doctrine when the companies asked for help in breaking into the few British preserves in the Middle East where interesting concessions were still obtainable. On the other hand, the companies and the State Department did not always see eye to eye elsewhere. In 1932, for instance, the Department stopped Jersey's subsidiary in Peru, the International Petroleum Company, from providing oil worth the equivalent of two and a half years advance taxes to the Peruvian government which needed the money to buy arms from the French in the face of an American arms embargo (Pinelo, 1973, p. 38). Under FDR the State Department remained unpredictable on issues concerning the fate of oil company property. When, in 1937, Jersey was expropriated in Bolivia, the State Department recommended that the company should resort to the Bolivian courts and it was not until 1940 that the USA moved to help the company get some compensation (Wilkins, 1974, p. 225). In the case of the Mexican expropriation, the State Department again stood aside after the Treasury Department tried economic pressure by suspending the silver purchase agreement. The State Department did finally protest to the Mexicans, but admitted the right of Mexico to expropriate with an obligation to compensate victims. Roosevelt specifically warned that there was to be no revolution and the US ambassador was forthright in criticising the companies and softening despatches to the Mexican government. The settlement of 1943 fixed compensation which was around one-eighth of what the companies had originally demanded (Turner, 1973, p. 111).

There can be little doubt that the personality of FDR and the policies of his New Deal administration were tinged with considerable anti-oil-company sentiment. When this was allied with his Good Neighbor policy, pledging the USA to a policy of non-intervention in the affairs of its southern neighbours, the companies could not count on State Department support as they had been able to do in Mexico until the early 1930s (O'Connor, 1972, pp. 270–87). At the same time that the USA was, for economic reasons, focusing on its domestic oil industry, the Democrat administration was more suspicious of big business than any government since Teddy Roosevelt's presidency. A coincidence of political and economic factors led the US government into a more ambivalent role towards its oil interests.

There was thus a combination of factors which reduced the involvement of parent governments in the doings of the majors after Achnacarry. The key Middle Eastern concession agreements had all been signed by the mid-1930s, so neither British nor American companies needed to call for diplomatic support in concession hunting. The incursion of new competitors was reduced by the Great Depression's impact. In addition, the anti-business ideology of the New Deal meant that US companies could not count on the sympathy of the State Department – and vice versa. In 1936, the US Secretary of State, Cordell Hull, tried imposing a 'moral embargo' on the export of essential raw materials, including oil, to Italy after the invasion of Abyssinia – but to no effect. American oil companies expanded their sales to Italy and Hull was forced to confide in his memoirs that 'a moral embargo is effective only as to persons who are moral' (Scott, 1973, pp. 352–3). An attempt by FDR and Hull to make this 'moral' embargo legal foundered in an isolationist Congress.

The failure of the Italian oil embargo is characteristic of the political distance between the oil companies and their parent foreign policy establishments during the mid-1930s, which was not reduced by the growing probability of a world war. The British had created an Oil Fuel Board (later, the Oil Board) in 1925 to keep the question of oil and tanker requirements under constant review in anticipation of future wars, but the oil companies were not given a formal role in planning until after the outbreak of the war in Europe (Payton-Smith, 1971, pp. 39–42). In an age in which policy makers were divided about how to face the threat from Hitler, the oil companies had their divisions too. Texaco, despite the US neutrality laws, supplied Franco during the Spanish Civil War. Later it was to supply the Germans during the early days of the Second World War. Sir Henri Deterding was increasingly identified with the Nazis and was eased out of Shell. Once war started in Europe, the US majors were faced with the dilemma of being connected with both sides of the hostilities. Jersey continued to exchange technical information with the German chemical combine, I. G. Farben, under their prewar contractual arrangements and was to pay for this action when the USA joined the war by becoming the prime target of two well-publicised public investigations.

In the Far East there was less ambivalence. The oil companies and the State Department had interacted in response to Japanese expansion since the mid-1930s. In 1934, Stanvac and Shell tried to get the British and American governments to mastermind a crude oil embargo against Japan, in retaliation for its creation of a Japanese oil monopoly in Manchuria, and on its insistence that foreign oil companies in

Japan should build up strategic oil stocks. Cordell Hull turned this proposal down, choosing to put the wider interests of USA-Japanese relations ahead of the commercial interests of the oil industry. However, despite this rebuff, the incident masks an extremely complex relationship between companies and the USA, British and Dutch governments up to the formal declarations of war. Both the diplomats and oil men saw Japanese economic nationalism as the harbinger of Japanese militarism, and, as Japan's territorial ambitions became more and more obvious, it was the diplomats who took charge. In one of its few major exercises in oil diplomacy, the Dutch government arranged with various companies in the Dutch East Indies for the destruction of oil facilities in case of a Japanese invasion. In the American case, by 1940–1, there were so many agencies involved in setting the relevant policies that the dividing line between commercial and strategic interest was totally blurred (Wilkins, 1974, pp. 230–3, 250; Anderson, 1975, p. 197).

OTHER CONSUMER GOVERNMENTS

Although the victors of the First World War were able to impose their will on the world oil industry until 1939, other Western governments developed relationships with the industry which, in some ways, anticipated the future structure. The creation of national champions, the protection of domestic markets from the full blast of the majors' competition and the search for security of supplies were all found in the 1930s as well as in the late 1970s.

As a marginal parent government, the French experience was an interesting contrast to that of the Anglo-Americans. Emerging from the 1914–18 War with a clear perception of vulnerability, the French did badly out of the negotiations over Iraqi oil. Worries about their dependence on imported energy were superimposed upon a wider despondency as political impotence was translated into a wider loss of empire. Not only did they allow the British to go back on an initial agreement that they should control the potential oil-bearing areas of Mesopotamia, but they were relatively powerless bystanders as the Americans muscled their way in during the 1920s. However they did win a stake in IPC, but this was not sufficient to supply the domestic French market. It was decided to create a French national champion rather than rely on Shell. By the end of the 1920s the Compagnie Française des Pétroles (CFP) was given a brief which stressed its role in protecting national interests as well as making profits. Primarily a company put together by private interests, the French government came to take a 35 per cent shareholding and a 40 per cent voting right (Rondot, 1962, pp. 31–7), the kind of private-public mix which had been pioneered by Anglo-Persian in Britain.

At the same time, the government established a monopoly on the import of oil into French territory and then delegated this monopoly to interested enterprises. The resultant system of import authorisations has survived into the 1970s and is still used to give French interests the maximum possible stake in the French market, thus limiting, but not destroying, the role of the majors. A further action discriminated in favour of export refineries by using authorisations to limit the imports of oil products (Mendershausen, 1976, p. 27). All told, this was an array of interventionist measures imitated by none of the other three parent governments. The USA, Britain and Holland have generally allowed their oil markets to function with the minimum of government interference, although there has been far more government activity in the domestic US oil and gas market than official ideology would have one believe. In contrast the French, since 1929, have pursued a notably *dirigiste* policy, whereby access to the market has been limited to a few companies which have been assigned market shares, have been favoured if they were French and have been preferred if they had access to crude oil from within the French Empire.

As the losers in the First World War, the Germans had more pressing concerns to worry about than the oil industry, such as the securing of the coal industry in the Saar. As it happened, there was some domestic oil production which removed some of the pressures to try to re-establish Germany in areas like the Middle East. However, awareness of the paucity of secure supplies meant that there was quite a lot of emphasis on producing petrol from coal (a technology which would get a new impetus in the late 1970s). Local affiliates of the majors were forced by law to accept that their first responsibility was to the German state and not to their shareholders (Tugendhat and Hamilton, 1975, p. 110).

Italian policies toward the industry showed more of the corporatist philosophy that was an integral part of European fascism. Azienda Generale Italiana Petroli (AGIP) was created in 1926 to look for oil in Italy and the Middle East, but it had its problems and had to sell off its interest in an Iraqi concession to help Mussolini finance his war against Abyssinia. Some oil was found in Albania by the state railways, and AGIP was able to take over some Romanian production. On the domestic front, the Licensing Act of 1934 was designed to encourage the construction of refineries in Italy and a tax was imposed on imported products. As in France, the national champion was not strong enough to take over more than a portion of the domestic market. However, by 1939, Italian policy had been successful enough to restrict the sales of foreign companies to around 60 per cent (FEA, 1975, p. 85; Tugendhat and Hamilton, 1975, p. 110), and the postwar policy of Mattei's Ente Nazionale Idrocarburi (ENI), which came to administer AGIP, was very much in the vein of prewar Italian economic nationalism.

The Spanish variant of the corporatist approach was to set up the state-controlled Compania Arrendataria del Monopolio de Petroleos (CAMPSA) which, from 1927, was the only permitted marketer of oil products within the country. This did not stop the international industry from being involved in refineries, but in 1939 the Ministry of Industry adopted regulations limiting foreign participation in the Spanish industry to 25 per cent (FEA, 1975, p. 165). On the other side of the globe, the Japanese also encouraged the founding of locally owned refining and distribution companies, sought to reduce their dependence on crude imports from the USA and the Dutch East Indies, and tried to insist, by the 1934 Petroleum Bill, that foreign companies should hold six months of stocks as a way of building up national oil reserves. The majors managed to circumvent the latter requirement through Shell and Stanvac's offer, with State Department approval, of the results of their research into the infant hydrogenation technology for producing oil from coal. The companies had a harder time in Manchuria, where Texaco sold out, and Stanvac and Shell tried to sell out, because of the oil monopoly set up there by Japan's puppet state (Wilkins, 1974, pp. 230–3).

Taken in conjunction with the widespread economic nationalism then found in Latin America, the *laissez-faire* approach of the Dutch, British and Americans appears to be the exception during this period. In fact, the official British backing given to Anglo-Persian indicates that even the *laissez-faire* governments had significant views in common with more protectionist governments. The difference was that American companies had an overwhelming financial and technological advantage stemming from their vast markets and production in America. Shell owed its strength to long-established international production and exploration expertise as well as its skill in tying its fortunes to those of the British Empire while Anglo-Persian built on its links with this empire, the Royal Navy and its control of one of the world's largest sources of oil.

National champions of other countries were trying to expand under the protection of governments with insufficient international clout. The majors were able to preserve significant market shares for themselves even in the most xenophobic countries, although there are signs that Italy and Spain might have made more spectacular incursions into the majors' preserves had not the Second World War come along. The defeat of the Axis powers meant that the dominant position of the majors was projected into the postwar decades. But the nationalism of the 1930s had not been destroyed and would re-emerge when governments of less developed countries started to become generally involved in the running of their economies. But by then the Anglo-

American hold on the producing world had started to slip, so their new national champions had a better chance of establishing themselves.

NOTES

1 In this deal the leading companies interested in Middle Eastern oil agreed that no one would undertake independent operations in an area roughly equivalent to the former Ottoman Empire unless the others agreed (Longrigg, 1968, pp. 68–70; Penrose, 1968, p. 94). The practical effect was to slow down the development of new sources of oil in areas such as Saudi Arabia and Bahrain, which fell within the Red Line drawn by Gulbenkian to show his understanding of pre-1914 Ottoman boundaries.

2 Gulbenkian, an Ottoman subject of Armenian nationality, was eventually to settle in Portugal. In oil affairs, he tended to side with the French.

3 The USA had a similar scare in 1942–3, Europe had one in 1956 after the Hartley Report to the OEEC and, to a less easily demonstrable extent, the Club of Rome's *Limits to Growth* report in 1972 encouraged many to feel that the world was running out of key resources such as oil.

4 Denny, 1930, pp. 3–26, gives interesting evidence of the inflamed state of some public opinion during the Coolidge era.

5 See Hexner, 1946, for other examples.

3

Majors and Parents 1939–69: The Second Industry Restructuring

The Second World War brought the parent governments back into the oil scene with a vengeance, and they remained actively involved as the changed realities of the postwar international power structure were translated into readjustments in the oil industry. The changes were not as drastic as those involved in the years following the First World War, but they established a structure which lasted with minimal modification until the OPEC challenge in the 1970s.

Oil was of crucial importance in the war. As in 1914, the Germans found themselves in permanent danger of being starved of this mineral, and tailored their military strategy accordingly. Hence in the early days of the war, there was emphasis on lightning attacks and German strategists remained interested in the oil-producing areas of Romania, Russia and the Middle East (Tugendhat and Hamilton, 1975, pp. 112–18). The Japanese were similarly vulnerable and concerned with reaching the oilfields of the Dutch East Indies and Burma. The allies, on the other hand, faced the problem of maintaining the flow of existing supplies of oil in the face of a startlingly effective German U-boat offensive. The British gave the oil companies a great deal of responsibility to meet this challenge. Political guidance came from the Oil Control Board, which was a subcommittee of the War Cabinet, but although this was originally conceived as an executive organ, it soon became clear that its members had neither the time nor the expert knowledge to make it work as such. It met about once a month during the first two and a half years of the war, and once every two months after that, leaving day-to-day operations very much to the oil companies, which primarily meant Shell, Jersey and Anglo-Iranian. They had started contingency planning at the time of the 1938 Munich crisis and were given government backing to set up the

Petroleum Board which combined all marketing and distribution operations into one centralised operation run from Shell-Mex House. As the war developed, these companies were formally given the responsibility of programming oil supplies throughout the empire. This enhanced role was reflected in the creation of the Trade Control Committee which, from the spring of 1940, was the supreme directorate of the 'sterling' companies, controlling overseas operations as well as distribution within Britain. Meeting three or four times a week and, for the first two years, with a government representative as chairman, this was the heart of the British oil industry's contribution to the war effort. The extensive delegation of responsibility to the companies was to prove characteristic of the relatively free-and-easy relationship between the British oil industry, civil service and political world, which was to be maintained at least into the 1960s.

Once the Americans entered the war, the centre of gravity of the industry's effort shifted westward across the Atlantic. The US approach to oil logistics was less relaxed than the British, with the military being extremely cagey about collaborating with civilian oil executives. Attempts to set up one over-arching co-ordinating body, the Combined Petroleum Board, foundered on Britain's uneasiness about putting what was, in effect, their combined civil and military supply operation under a committee whose US members would be entirely military. The British won this particular battle. It was accepted that the problems of oil supply were too complex and specialised to be left in purely military hands, and that the requisite co-ordination could be achieved through informal liaison with national planning bodies and through the creation of *ad hoc* Anglo-American committees set up for specific purposes. A mixture of British officials and company men based themselves in New York and Washington to consult with their American counterparts on problems of global production and distribution, while a lesser number of Americans likewise based themselves in London. Oil companies based in North America were banded together into the Foreign Operations Committee which reported to the US Petroleum Administrator for War. This committee was charged with working out oil supply arrangements for foreign areas not controlled by the enemy and was the equivalent of the London-based Trade Control Committee. The companies were officially recognised as key actors on either side of the Atlantic, and the complex web of transatlantic, industry-government relationships was given a final embellishment as far as the companies were concerned by the creation of this twin committee system, one for the 'sterling' companies and the other for their 'dollar' opposite numbers. Whenever issues called for co-operation between companies on either side of the Atlantic, both the London and New York committees had to approve (Payton-Smith, 1971; Church Report, 1974, *IPC*, p. 13).

Official histories do not show how much these war committees were related to prewar attempts by the companies to create an international cartel. The Church Sub-Committee on Multinational Corporations claimed that the personnel on the American side were substantially similar to the membership of earlier steering groups, and that the allocations of supplies made by the Foreign Operations Committee 'were identical in character with those found in the files of the companies for operations prior to Pearl Harbour' (Church Report, 1974, *IPC*, p. 13). There is also evidence that even the British government felt a bit uneasy. The government official who attended meetings of the Trade Control Committee as an official observer gave up the practice in mid-1942 as it was apparently felt that his regular attendance was giving an aura of official respectability to the committee (Payton-Smith, 1971). However, whatever the hesitations of officials, governments had to work through the oil companies as these alone possessed the kind of specialist knowledge needed for running day-to-day operations as well as for making many long-term strategic decisions. The pattern of functional industry committees under a higher strategic body to consider wider political interests has seemed both natural and inevitable to the oil industry not only in wartime but also during the Suez crises, the 1973 oil embargo and in relations with the International Energy Agency. This was true not only in Britain but also in the USA, despite the fact that the companies there were confronted with a suspicious military establishment and a Democratic administration.

One bizarre example of this tension occurred during the war when Socal and Texaco became afraid that the British were wooing the sympathy of King Abdul-Aziz and that their holdings in Saudi Arabia might therefore be in jeopardy. Lend Lease aid was being sent to Saudi Arabia from 1941 onwards via the British, despite the fact that the Saudis hardly counted as 'democratic allies', and the British were naturally wringing the maximum of goodwill from the process. So worried were the two American companies that Texaco's chairman offered to set aside a separate petroleum reserve from which US government needs could be supplied at preferential prices, providing that the USA sent Lend Lease directly to the Saudis. It was clear that CASOC (the joint venture of Socal and Texaco which was to become Aramco) had only survived since 1941 with heavy US diplomatic and financial assistance. Although Washington saw the wisdom of declaring the defence of Saudi Arabia 'vital to the defence of the United States' in order that the Saudis might qualify for Lend Lease, there was a serious debate about taking up Texaco's offer. The Petroleum Administrator for War, Harold Ickes, was a director of the Petroleum Reserves Corporation (PRC), created in 1943, along with the Secretaries for State, War and the Navy. The PRC started negotiating with

Texaco and Socal not just for a separate petroleum reserve but for 100 per cent ownership of their joint holding in Saudi Arabia. The companies were seriously taken aback by this sudden demand, which when far beyond the initial trade-off they had been willing to consider to get government help, and negotiations were soon suspended. The PRC was then subjected to a major onslaught from the industry which drew a parellel between the proposed US government stake in CASOC with nazi and fascist philosophies. The State Department decided that foreign oil was equally accessible regardless of whether it was in the hands of private American companies or in those of a state corporation and the PRC was allowed to lapse in 1944 after one further initiative aimed at building a pipeline from the Persian Gulf to the eastern Mediterranean (Church Report, 1975, pp. 38–41; Sampson, 1975, pp. 94–9). This episode was an exception to the general pattern in which the ability of the majors to keep themselves formally independent of Washington, only calling for assistance as the occasion demanded, temporarily slipped. However, the ability of CASOC to get the USA to finance Saudi Arabia in exchange for a promise of a petroleum reserve which was then withdrawn is one of the two or three most classic cases in which oil companies manipulated parent governments for their own purposes.

The fact that Saudi Arabia could become a bone of contention between companies and the US government was symptomatic of the growth of general interest in the Middle East. Up to the early 1940s, not enough oil was being produced within the area (except in Iran) to make much impact. What little there was (Iraq was the only other significant producer during the 1930s) was too expensive to be worth transporting to North American markets. However, prewar discoveries in Kuwait, Bahrain and Saudi Arabia had alerted the companies to the potential of the area and the US government came to accept the conclusion of the geologists sent out in 1943 on behalf of the PRC that: 'The center of gravity of world oil production is shifting from the Gulf-Caribbean areas to the Middle East, to the Persian Gulf area, and is likely to continue to shift until it is firmly established in that area' (Hartshorn, 1967, p. 319). The only trouble was that the British controlled 81 per cent of Middle Eastern oil production in 1943, against America's 14 per cent – a commercial division which did not correspond to the new realities of the emerging world power structure (Church Report, 1974, pp. 41–2). As a result, beneath the civilities of Allied co-operation in the war effort, the British and US governments were fighting over an American attempt to re-draw the structure of the Middle Eastern oil industry in a form more equitable to the USA. In the postwar era, this was to involve the French as well, as potential losers in this process along with the British. This second major restructuring of the international industry was not completed until 1955, when the last bits and pieces were added to the settlement of the Iranian dispute with what was now BP. By then, the USA had ensured that the expansion of Saudi production was to take place under American corporate control and that the British monopoly of Iranian oil was broken. The balance of corporate control in the area had been turned upside down, and the political future of local authorities in the producing countries was inevitably linked to their relationship with the USA, parent government of most key companies.

The expansion of US influence was a case of government and corporate interests marching hand in hand. The fear, in the winter of 1942–3, that the country was running out of oil (planes were short of 100 octane gasoline, and domestic reserves were being drawn down faster than new reserves were being discovered) fed the belief that the USA should be involved directly in the Middle East (Church Report, 1975, p. 38). Hence, in the same surge of inter-Allied diplomacy which produced the Bretton Woods Agreement, the United Nations and a structure for the postwar international airline industry, negotiations started which culminated in the Anglo-American Oil Agreement of 1944, which was an American initiative with apparently two aims. First, it was yet another attempt to reassert the Open Door doctrine against British control in the Middle East. Secondly, it was an effort to find an 'orderly' approach to the potential future glut of oil which seemed likely, given what was then known about the scale of discoveries in the Middle East. The key was to be an International Petroleum Commission, which was to analyse supply-demand issues and short-term problems arising in production, transportation and processing. It was to report its conclusions to each of the governments 'and to recommend to both governments such action as may appear appropriate' (Church Report, 1975, p. 43).

Although somewhat of an aberration in the history of US oil policy, this agreement was very much a child of its time; 1944 was also the year when the Chicago Conference was held to decide the structure of the airline industry, the Americans hoping to get agreement for a system which would give its experienced and well-equipped carriers the maximum opportunities in the postwar world. When the Americans did not get the agreement they wanted at this conference they entered the 1946 Bermuda Agreement with the British, who also had great interest in a solution. With both airlines and oil, the USA sought international solutions in which the freedom of US corporate interests would be extended by intergovernmental agreement. In both cases, Britain was the most important power with which to do a deal (Straszheim, 1969, pp. 32–3). However, the oil companies were much less resigned to working under a governmental framework than the airlines and saw to it that the Anglo-American Oil Agreement of 1944 was consigned to the oblivion awaiting all international agreements (such as the 1948 International Trade Organisation) which fail to get ratified by the US Senate (Caroe, 1951, pp. 112–17).

Meanwhile, battlelines between the defending British and attacking US commercial interests were being drawn. Both sides knew that the development of key wartime projects such as pipelines and refineries would confer significant postwar commercial benefits. The Roosevelt administration eschewed projects which might build up the British position in the Middle East and the British found it necessary to bypass the Oil Control Board (the Cabinet subcommittee which oversaw oil matters) during the last couple of years of the war, since there were American observers on it, and there were too many issues involved which bore on the future of postwar Anglo-American oil relations (Elwell-Sutton, 1955, pp. 140–1; Payton-Smith, 1971, p. 478; Church Report, 1975, p. 42).

SCRAPPING OF THE RED LINE AGREEMENT

The first postwar battle of note involved Jersey Standard's decision that, along with Socony, it had to break into the Saudi oil preserve. This was a complicated affair, involving intergovernmental diplomacy. The pressure for change came from Jersey which felt threatened by Socal's and Texaco's hold over supplies of reserves which were clearly very extensive and very cheap. As its production was increasingly absorbed by North American markets, Jersey found itself with a perilously weak productive capacity with which to supply its traditional share of European and Japanese markets and feared that Socal and Texaco would use Saudi oil to break into them. In 1946, Jersey, in conjunction with Socony, decided to find a way into Saudi Arabia, despite the fact that its IPC membership theoretically meant that it was bound by the Red Line Agreement which barred IPC participants from independently seeking concessions in the area in which Saudi Arabia was located. The CASOC partners were quite easily won round to Jersey's proposal, though there was some debate within Socal as to whether the company should not 'go competitive' and use its hold on Saudi crude oil to increase substantially its share of the world market. There were, however, two major problems. First, the proposed deal was an anti-competitive act, especially as Aramco's financial structure was so constructed as to discourage ultimate market price competition between the various partners. The Justice Department was consulted and allowed the deal to go ahead in 1947 on the grounds of national security, defending its decision on the grounds that:

> [there is] a predilection on the part of the State Department to consider international oil as a part of foreign policy. The Depart-

> ment of the Interior, Department of National Defense and National Security Resources Board are also concerned with the shortage of oil reserves in this country. It is doubtful that they would look with favor upon any anti-trust case which might affect the American positions in foreign oil . . . (Church Report, 1975, pp. 45–55; Sampson, 1975, pp. 99–104)[1]

There was also the problem of getting the British and French to permit Jersey and Socony to slide out of the restrictive Red Line Agreement. The State Department went on the offensive in a repeat of its performance of the 1920s when the US government had won the US companies their original stake in Iraq. The British were not too difficult, since both Shell and Anglo-Iranian realised that the legality of the Red Line Agreement was now in question and were worried that the French might bring the whole affair into the open. With State Department blessing, Anglo-Iranian was won round by a long-term sales agreement in which Jersey and Socony agreed to take large volumes of its crude oil for twenty years (Church Report, 1975, pp. 50–2). The French were tougher opponents, with the aggrieved CFP and Gulbenkian doing everything they could to ensure that they were invited into Saudi Arabia through the old 1928 agreement which the French government claimed, in a warning to the US government, it considered intergovernmental. The State Department could see that life would indeed be simpler if all the IPC partners were allowed into Saudi Arabia, but was persuaded by the American companies that the Saudis would not accept this, and that a compromise could be reached whereby CFP and Gulbenkian would be given more than their official share of Iraqi oil, in return for permitting the dissolution of the Red Line Agreement. This was indeed so, and the US companies were able to buy the French off, just as they had bought off the British. However, it was a deal which left the French more convinced than ever of how badly the Anglo-Saxons behave if given a chance.

Iranian Nationalisation

The final postwar expansion of US interest in the Middle East occurred in Iran. Unlike the Saudi case, American involvement in Iran was not the result of commercial ambitions but appeared as a necessary response to the dangerous political vacuum created by Britain's inability to come to some mutually agreeable compromise with Iranian nationalism. The involvement of parent governments at the highest level was particularly marked in this case. Though inter-company diplomacy was important in devising a formula which would satisfy the Iranians, governments were involved, first in ensuring a government which would be able to accept such a formula (achieved via the CIA-backed overthrow of Mussadiq), and secondly, in sorting out which companies would be allowed in. There was never one single British position, with Palmerstonians like Foreign Secretary Herbert Morrison balanced by the cooler counsels of Prime Minister Attlee and much of the Foreign Office. The Labour government which first dealt with the crisis was succeeded by Winston Churchill's Conservative administration, in which Churchill saw the chance of reliving one of the more notable moments of his early career when he had helped establish Anglo-Persian as one of the great oil companies of the world. British responses appeared to many as blimpish and narrow-minded, oblivious to the lesson of India's independence that all countries in Britain's formal and informal empires (Iran at the time fell into the latter) would one day want to assert themselves in the political and economic spheres. Sampson argues that the British were singularly unprepared for the crisis as the information they received filtered to London through Anglo-Iranian (1975, pp. 118–21).[2]

The first official British action was the classic response of the affronted power – the dispatch of a warship. Then the government settled down behind the oil companies' preferred strategy of boycotting Iranian oil and RAF planes forced a ship which was carrying some of the disputed oil into the British colony of Aden. Meanwhile, the State Department indicated that US companies would not help the Iranian government in any way (Sampson, 1975, pp. 118–20). The British, under Churchill's administration, were certainly keen that Mussadiq be overthrown and Sampson claims that the original idea for a Western-sponsored coup came from, and was finally sanctioned by, the British (though the Americans had their own wider reasons for giving the CIA the go-ahead) (1975, pp. 126–7). After Mussadiq's fall there were a series of talks between the State Department and the Foreign Office, aimed at finding a way of co-ordinating the efforts of companies and governments to counter the growing nationalism which could be sensed in the Middle East. The British pointed to the growing dependence of Western Europe on Middle Eastern oil and argued that they were not just concerned with protecting the profits of British companies. They even floated the idea of a revamped International Petroleum Commission, as suggested in the Anglo-American Oil Agreement of 1944. This time round, the commission would not only hold bi-annual co-ordination meetings between American and British officials, but would embrace representatives of both producing and consuming countries – a concept well ahead of the times (Church Report, 1975, pp. 64–5). However, there were a number of American objections to this idea, and whatever initiative the British had faded in front of the State Department, now the dominant political force in oil negotiations.

Throughout this crisis, the Americans had been able to stand back and take a somewhat broader view of the issues at stake. From the start, they were critical of British policy, particularly as they had already steered their own companies through demands for fifty-fifty taxation in Venezuela (1948) and Saudi Arabia (1950), avoiding the kind of diplomatic catastrophe with which the British were confronted in Iran. However, top US policy makers, from Truman and Secretary of State Dean Acheson downwards, were aware of the wider issues, such as the relatively weak state of the European, especially British, economies, and the danger that the Iranians might be driven into the arms of the USSR by any long-drawn-out campaign by the British. These were relatively early days of the Cold War, and Iran adjoined the Soviet Union, which had not been averse, in the past, to staking a claim to some parts of northern Iran and had been somewhat reluctant to remove troops stationed in Iran during the Second World War when peace was declared (Landis, 1973, pp. 34–6).

As the dispute dragged on, it became increasingly clear that some formula had to be found in which Anglo-Iranian's role would be greatly reduced. It was essential that whatever post-Mussadiq government emerged would be able to sell its oil in world markets and feeling against the British company ran too strong for any Iranian leader to accept a continuation of British dominance of Iranian oil exports. Other companies had to be brought in. Shell could obviously absorb some of this oil, but it was increasingly obvious that the US majors would have to come in as well. The eventual solution was a widening of the consortium to include various majors and a group of American independent companies.[3] Although this appeared to be an extension of the drive by US companies to break into a hitherto British preserve, in fact the picture is more complicated. The US majors were not particularly enthusiastic about taking on commitments to market Iranian oil as they had been expanding production elsewhere in the Middle East and feared that they could only slot Iran's oil into world markets at the expense of oil from other Middle Eastern countries whose governments would not gladly accept a diminution of their production and thus exports. Also, they viewed with some unease the idea that Iran's treatment of Anglo-Iranian might be used as a precedent against other oil companies, including themselves, elsewhere, and were thus reluctant to see a company treated too roughly in any final settlement. This reluctance should not be taken too seriously, though company reservations were noted by official observers at the initial meeting held in December 1952 between the US majors and Secretary of State Dean Acheson (Church Report, 1974, *IPC*, p. 27).[4] But a refusal by the majors to co-operate with the State Department would have left a vacuum as far as Iranian oil was concerned, likely to be filled by smaller competitors like the US independents, which were looking for foreign oil and felt they could count on support from the antitrust authorities which had initiated a Grand Jury investigation into the alleged international petroleum cartel during the summer of 1952. In fact, the Department of Justice lost its initiative. The Departments of State, Defense and Interior argued within the National Security Council that the criminal antitrust case against the companies should be dropped and that a much less potent civil suit be initiated instead (Church Report, 1975, pp. 61–2). They won their argument. After President Truman decided that national security called for a gentler approach, he formally advised the Attorney-General to drop the criminal suit. Whatever the merits of this decision, it was to many a clear affirmation that the majors were acting as agents of the US authorities and thus could demand treatment which would not harm their reputations in the eyes of the rest of the world.

Early US efforts had been to restore Anglo-Iranian to its former status but, by October 1952, the State Department was convinced that the US majors had to be brought into Iran because, in Acheson's words:

> no one other than the majors and . . . [Anglo-Iranian] had sufficient tankers to move large volumes of oil . . . One of the concrete problems in securing a resumption of the flow of Iranian oil is to determine whom it is we can call on, and who is able, in fact, to move Iranian oil in the volume which is required to save Iran. The independents are not in a position to give us any real help. (Church Report, 1975, p. 60)

The five US majors entered into formal discussions with the Secretary of State and it was thus assured that the ultimate solution would be built round them. The most which would be achieved by the Department of Justice was to see that a group of US independents were ultimately allowed entry into the new consortium.

Foreign Tax Credit Issue

The State Department's acceptance that the majors could be useful agents in providing a viable economic solution for Iran was not unique. They had played a similar role two years previously when Middle Eastern governments, particularly the Saudi, became interested in the fifty-fifty tax system which the Venezuelans had recently achieved (Wilkins, 1974, p. 321).[5] The Saudis became increasingly impatient with Aramco, which, in its turn, became increasingly worried about the growing anti-Aramco, anti-American tone in Saudi Arabia. The outbreak of the Korean War made the stability of the Middle East of even greater importance than before. Washington therefore considered a number of ways of getting more money to the

Saudis. It was argued quite seriously within State Department circles that Aramco should hand back a large portion of its concession to the Saudi government which could then raise income by offering it to other oil companies – but the Aramco partners were understandably something less than enthusiastic about this idea. An alternative would have been for Aramco to pay a considerably higher royalty to the Saudis, but this was a cost which would be passed on ultimately to the consumers, then primarily the Europeans, whose economies were just being restored under the Economic Co-operation Administration set up to administer Marshall Plan aid. High oil prices would be a serious setback to this programme which was seen as so crucial in strengthening the anti-communist forces in Europe.

Another solution was to allow Saudi authorities to tax Aramco's profits and, with US Treasury permission, have this tax offset against taxes Aramco were paying in the United States, thus leaving the price of Saudi oil to the customer untouched and transferring the burden of increasing Saudi income to the US taxpayer. This was the solution which was finally adopted after a debate in the National Security Council. The result was: 'In 1950, Aramco paid the United States $50 million in income taxes; in 1951, the company paid only $6 million. Conversely, payments to the Saudis increased from $66 million in 1950 to almost $110 million in 1951' (Church Report, 1975, p. 85).

In later years, this decision received a great deal of hostile attention, and Senator Church's Senate Sub-Committee devoted considerable time to unravelling its history (Church Report, 1974, pt 4, pp. 14*ff*., 85*ff*., 114*ff*.; Church Report, 1975, pt 8, pp. 350–5; Blair, 1976, pp. 195–203). In fact, the principle of foreign tax credits had been established by the American authorities as early as 1918, and the pattern of giving tax concessions to companies dealing with 'favoured' sections of the world had been established even earlier for those doing business with China and elsewhere (Wilkins, 1974, p. 262). Given the fact that, in 1950, the posted price[6] declared by the oil companies was basically the market rate, so Aramco's declared profits were not arbitrarily determined, there was nothing particularly unusual about the solution (Church Report, 1974, pt 4, p. 89) in Saudi Arabia and according to Ambassador McGhee, the solution was widely known at the time (Church Report, 1975, p. 84). In Saudi Arabia, as in Iran, foreign policy makers realised that oil companies were useful agents. This was expressed clearly in a key background paper prepared in September 1950 for a State Department meeting with oil executives to discuss the issue of rising Middle Eastern demands:

> to maintain and protect the West's oil position in the Middle East and to attain overall US Middle East policy objectives, . . . increased

efforts must be made on every available front to prove that the American way of life and the American meaning of democracy can be made to have meaning for people living in the Middle East and can be shown to offer more rewards during their generation than those offered by others. In these efforts, oil companies are playing an essential role. Oil companies occupy . . . not only an integral part of Middle East affairs but present the West with its broadest contact with local peoples at the lowest level. Thus, to a large extent the maintenance and protection of Western oil interests in the Middle East depends on the attainment of overall US policy objectives for the area and similarly to a large extent the attainment of these objectives depends on the manner in which the oil operations are carried out. (Church Report, 1974, pt 7, pp. 128–9)

In statements such as these, the State Deparment voiced the feeling which dominated Washington in the tense days of the late 1940s and early 1950s – the international oil companies were too useful in the struggle to preserve the 'Free World' from communism to be exposed to disruption by antitrust authorities or to interference by an overzealous defence of the interests of US taxpayers. The interest of national security came before all else.

PARENT GOVERNMENTS WITHDRAW AGAIN

The settlement of the Iranian oil crisis in 1954 marked the end of a particularly turbulent period in the history of the oil industry. The subsequent sixteen years were to be very different, with the governments of virtually all the industrialised countries relegating international oil politics well down their list of priorities. Certainly, there were important crises surrounding the 1956 and 1967 Suez Wars, and there were disputes with particularly militant governments like those of Iraq and Peru, but in general foreign policy makers could forget about the industry. The USA, in particular, turned in on itself, shutting its domestic market off from rapidly increasing imports of cheap foreign oil. Most governments were more concerned with running down their coal industries than with the security of oil supplies (with the notable exception of France, whose struggle against Algerian independence was heavily influenced by the existence of oil in the Sahara). However, the general lack of government interest in the international oil industry does not mean there was a complete break with the traditions of previous decades. In the few times of genuine crisis, the companies and parent governments still worked hand in glove and, as the years went by, the companies slowly expanded their claim to be general agents of the industrialised world, though non-parent governments showed unease about relying overly on the majors and experimented with alternatives.

The key factor determining the complacency of industrialised governments was the fact that oil production from non-communist sources was in comfortable oversupply for most of this period, growing at 7·2 per cent per annum from 1957 to 1970. At the same time, proved oil reserves grew only slightly less rapidly, with well over two-thirds of the new discoveries being in the Middle East and North Africa, where production costs were extremely low. In the early 1960s, the investment needed to bring a daily barrel of crude oil on stream was less than 5 per cent of that needed to produce an equivalent barrel in the United States (Adelman, 1972, p. 76; Darmstadter and Landsberg, 1975, pp. 32–3). As far as the oil companies were concerned, the problem was not finding sources of crude oil, but of limiting production without offending the various host governments.

In so far as the industrialised governments thought about oil in a serious way, it was seen as a threat to domestic energy production. For the British and West Germans in particular (and, to a lesser extent, for the Japanese and other Europeans), this primarily meant the coal industry which was shielded by a variety of measures designed to preserve it where possible, or at least avoid any catastrophically fast rundown of such a politically powerful industry. These measures included directives to both state-owned and private electricity power producers, restraining their freedom to use oil as a fuel, as well as taxes on oil products (Adelman, 1972, p. 227; Odell, 1972, p. 94). Needless to say, it proved impossible to reverse the tide whereby the economies of the industrialised nations moved from a coal to an oil base. In 1950, coal provided 56 per cent of world energy consumption, compared with 29 per cent by oil: by 1970, the positions were reversed with coal's share dropping to 31 per cent, while oil's rose to 44 per cent (Darmstadter and Landsberg, 1975, p. 19).

The American position was somewhat more complicated, since a major domestic oil industry existed alongside significant coal operations. The interplay of these industrial sectors during the 1950s made for some fascinating politics. The late 1940s and early 1950s had seen an upsurge of interest by US independents in finding sources of crude oil abroad, partly because these would not be controlled by regulatory bodies like the Texas Railroad Commission which kept a tight hold on the domestic industry. By 1953, the coal industry was sufficiently worried about oil imports to sponsor a bill to impose import quotas on petroleum products. The coal lobby was joined by the domestic oil industry as most of the new sources of oil supply outside the USA were being developed at costs which would destroy the high-cost US industry if imports continued unchecked, particularly since the USA, in 1948, ceased to be a net exporter of oil. The majors were initially hostile or ambivalent to the demands for protection, since they had international interests to defend. Ultimately they came round to the idea of import quotas, since they were in a position to have their cake and eat it too. Protection of the US market would mean that their investment in domestic production would be safeguarded, while their Venezuelan and Middle Eastern production (particularly the latter) could be diverted to fast-growing European and Japanese markets. The result was a powerful combination of interests which led the Eisenhower administration to call for voluntary restraint on oil imports (1954 and 1958) and, after the voluntary approach had failed, to impose the 1959 mandatory oil quotas which lasted into the 1970s. Although intended as a protectionist measure, this 'drain-America-first' approach reduced American oil reserves at a time when the threat of host government militancy was minimal (worries about Soviet intentions were more understandable), so that when a real threat to the world oil industry emerged, the USA no longer had any spare capacity with which to counter it (Bauer *et al.*, 1963, pp. 363–75; Odell, 1972, pp. 29–30).

A strong belief in the industrialised world in the 1950s and 1960s was that the only real problem in the oil industry was the over-aggressive competition of oil companies with indigenous energy industries. The thought that oil supplies might be exhausted in the fore-seeable future or be manipulated by hostile governments did not cross the minds of many foreign policy makers. Their attitudes were best caught by the Robinson Report, published by the OEEC (the OECD's forerunner) in 1960. In contrast to a pessimistic study published four years earlier by the same organisation, the Robinson Report claimed that no persistent long-term shortage of energy was likely before 1975; that new supplies of oil from countries like Libya meant that sources of oil were diversifying; that the oil companies' internally generated funds would be enough to provide most of the needed investment. Admittedly there was some discussion in this report of the wisdom of increasing the size of stocks held in Europe but, in general, its message was clear: oil supplies could be safely left to the oil companies, while importing governments should concentrate more on the smooth running down of the uncompetitive coal industry.

This complacency had been strengthened by the 1956 Suez crisis and was reinforced by the oil industry's ability to ride the 1967 crisis with even greater ease. Above all, the lesson of these two crises seemed to be that the oil companies were able to rearrange world oil-flows to minimise the economic damage threatened by an event like the closure of the Suez Canal and were thus in a position to emasculate any attempt to use oil as a weapon against the importing nations.

Suez 1956

The first crisis came when the 1956 Suez War led to the disruption of traffic through the canal. Although imported oil was not yet of paramount importance to Europe (West Germany was only 9 per cent dependent on it), the canal closure severely disrupted the oil industry's logistical system, and tankers had to be diverted to the much longer Cape route at a time when carrying capacity was fully stretched. By the beginning of November 1956, Europe was losing 1·8 million barrels a day, or two-thirds of its total supply, although some precautionary measures had been taken. Earlier in the year, when the chances of some Egyptian action against the canal had been seen to increase, the US authorities had taken the initial steps which were to lead to the creation of a group of company executives called the Middle East Emergency Committee (MEEC), which was to function in an emergency under government supervision.

The Europeans were slower off the mark, but, by the end of November 1956, had created the OEEC Petroleum Emergency Group (OPEG) and a number of domestic advisory committees in individual countries. The solution to Europe's problems was seen as the diversion of oil from the USA and Venezuela to compensate for the shortfall caused by tankers having to sail round the Cape of Good Hope. There was a delay in putting this plan into action in November when the USA temporarily suspended MEEC in order to pressure Britain and France into withdrawing their troops from Egypt, but by the end of the year American oil was being shipped across the Atlantic, and OPEG made the first allocations of emergency oil to OEEC member states in January 1957. By April, oil-flows were sufficiently back to normal for OPEG to make its final allocations and by May OPEG and MEEC were suspended. OPEG was formally disbanded in July, presumably as a gesture to the US antitrust authorities who had only reluctantly agreed to such extensive co-operation between the companies in the first place (OEEC, 1958, chs 2–5).

Along with the earlier 'defeat' of the Iranians, the success of this emergency operation was to dominate the industrial world's perceptions of the oil security issue for the coming decade and a half. The Americans had never been seriously affected by the crisis, since they were then minimally dependent on Middle Eastern imports, and in so far as their policy makers pondered over the problem, they supported the European conclusion that the companies could be trusted to maintain oil-flows in a time of crisis. The fact that this was the first time in which the OEEC was involved marked a difference from the days of the war when the British and Americans had controlled oil supplies very much between themselves. The OEEC started its somewhat unsystematic involvement with European oil affairs as early as

1948, when an oil committee (along with a coal one) was created to oversee the industry's regeneration under the Marshall Aid Programme. This committee was concerned with rationalising the oil-refining sector, and it played no role in the 1951–4 Iranian crisis.

The 1967 Suez crisis confirmed the industrialised governments in their complacency. Although the new closure of the Suez Canal caught them by surprise, this was no longer a major worry since more and more oil tankers were designed to sail round Southern Africa. Even the modified Arab embargo, in which the Saudis cut back production for the USA and the UK, was a dismal failure. The companies felt able to protect the West.

The formation of OPEC was greeted with general equanimity, although John J. McCloy (the spokesman for the majors in the USA) did approach the Department of Justice in case the companies might need antitrust clearance to co-operate in resisting OPEC pressure and the OECD's oil committee did make recommendations about stockpiling of oil supplies to provide increased security against politically inspired disruptions of supply (the EEC used this approach in 1964) (Hartshorn, 1967, pp. 292–3).

Of course, there were still oil matters which called for diplomatic activity on the part of parent governments, but none of these demanded the kind of attention granted through the 1940s and early 1950s. There was, for instance, the question of Soviet exports which seemed to threaten the price structure of the 'Free World' oil industry in the late 1950s and early 1960s. This problem originally manifested itself when ENI contracted to bring Soviet oil into Europe, but escalated as this oil found markets in West Germany, Sweden, Japan, France, Austria, Greece, Egypt, Cuba and Brazil (Tanzer, 1969, pp. 78–89). There were also significant diplomatic disputes with India, Cuba and Ceylon when these countries tried to force the majors to process Soviet oil in their refineries at the expense of crude oil from their own concessions. In the Cuban case, the dispute with the majors was a major factor in exacerbating relations between Castro and the US government (see pp. 73–5). In Ceylon the expropriation of the properties of Caltex, Stanvac and Shell led the USA to cut off aid in 1963 in accordance with the Hickenlooper Amendment which instructed US administrations to do this whenever foreign countries expropriated American property without appropriate compensation (Penrose, 1968, pp. 229–30).

The majors and the parent governments saw the issue of Soviet oil as both commercial and political. Jersey sent a letter to the State Department suggesting a boycott, the British refused to permit its importation and NATO embargoed exports of pipeline to the Soviet Union. The problem gradually faded away of its own accord when the Soviet Union realised that its surplus was not as large as it had assumed and thus reduced exports.

In general, this period from 1954 to the late 1960s posed only relatively isolated problems for parent governments. There were few signs that OPEC should be taken very seriously as a threat. The major dispute with a host government during this era was between Iraq and the Iraq Petroleum Company which ran on into the 1970s as pressures on the companies to resolve it were relatively weak, given the general glut of Middle Eastern crude oil and the feeling that Iraq's stand was not serving as a precedent for others. Apart from this case, it was the non-Middle Eastern hosts which gave the parent governments most trouble. Venezuela was consistently unhappy with the way the US import quotas worked against its interests, but this underlying grievance did not manifest itself in any major attack on the status of the majors. In Nigeria, Shell and BP got themselves caught in the middle of a civil war but this was more of a threat to the physical assets of the companies than to the concessionary principle which sustained the majors at that time. In Asia there was a bitter wrangle in Indonesia with the Sukarno government which passed legislation in 1963 to the effect that all foreign companies in future could only be contractors without concessionary rights. Although this was a radical attack on the oil industry's traditional structure, there was no sign that Sukarno's innovation would be adopted elsewhere.

Perhaps the most testing case for the US government during this period was the long-running dispute between the Peruvians and Jersey's subsidiary, the International Petroleum Company. The State Department became involved in the early 1960s and its manipulation of US aid programmes to Peru from 1964 to 1966 showed that it took the dispute seriously, though its efforts did not prevent the expropriation of the International Petroleum Company in 1968. Yet Peru, in fact, was a minor oil producer with little or no weight in oil circles. Washington apparently saw the dispute in the context of restraining Castro's influence in Latin America to the minimum, and Peru was perceived as a threat not so much to the general oil industry as to the interests of US foreign investors in this area.

Although the above incidents do suggest that pressures on the Western oil industry were growing during this period throughout the world, the key lay in the Middle East. If the importance of OPEC was discounted, as most Western observers of the time tended to do, the structure of the industry in this area appeared safe from radical attack. Iraq was the only oil producer in fundamental conflict with the majors.

On the other hand, Iranian oil had become an issue for the State Department in 1966, when the Shah launched a major campaign to persuade the Consortium to increase production at a rate which posed serious problems for at least three of the majors (Jersey, Socal and Texaco) which were both members of the Iranian Consortium and shareholders in Aramco (Mobil was sufficiently crude-short to want an increase in Iranian production). These companies were involved in a complex juggling act restricting their total offtake of Middle Eastern crude oil while trying to ensure that Iranian and Saudi production grew roughly in line. This became progressively more difficult as Iran's voracious appetite for income to be spent on arms and industrial investment increased disproportionately (Church Report, 1975, pp. 102–18). Inter-company conflicts were complex, as was the interplay between the relevant companies and the British and American governments which had different views on the extent to which the industry actually could lift Iranian crude oil at rates which would satisfy the Iranians. The US State Department was very much on the defensive, partly because the debate with Iran turned on secret agreements within the Consortium (and Aramco) on the terms by which member companies could 'overlift' – take more oil than the amount they were allowed to take at tax-paid cost by the Consortium's formula (the Aggregate Programmed Quantity system).[7] The State Department was well aware of the bitterness with which the shah would turn against the American companies should it become known that the Consortium's handling of this overlift question was less generous than the manner in which the Aramco formula handled it. In the words of consultant Walter Levy, these commercial arrangements were 'political dynamite' and in December 1966 the Department concluded that: 'These sensitive aspects make it desirable for the US government to limit its involvement in this problem unless urgent reasons arise for doing so' (Church Report, 1975, p. 111).

Twelve months later, State Department officials were urging Jersey and Mobil to liberalise Consortium overlifting arrangements as the Iranian government had learned the essential details of how these worked in both Iran and Saudi Arabia. Some changes were made, but not enough to stop Iranian pressures and the State Department continued to be worried about the problems the companies were creating. In March 1968, Under-Secretary of State Eugene Rostow met senior officials of the US companies involved in the Iranian Consortium. Normally, Rostow told the oil men, the Department kept out of commercial affairs, but the current dispute with Iran had national security implications which meant they should put national considerations above commercial ones. There was a danger that Russia might cut off Europe's oil supply and Iranian oil was important in reducing this risk. In the words of Iricon (the company representing US independents in the Consortium):

> State expects Russia to encourage and support the Arabs hoping to gain control of the Middle East oil that Europe is dependent on. As Iran is the strongest state in the area, it is very important to the US in maintaining influence.
>
> The overall situation is more serious even than last June. It could result in another blow-up and an oil boycott. Iran is not anxious to embargo oil production but would be pressured by the Arabs and might retaliate against the US companies for being reluctant to meet its demands for increased production and revenue. (Church Report, 1974, pt 7, p. 275)

The State Department was aware of the danger of a new embargo but did not feel that the US was vulnerable. Although oil company activities could be a positive embarrassment to Washington, discreet pressure on the US majors was still combined with concern lest American corporate interests be pushed aside by non-American competitors (Church Report, 1975, pp. 105–18). Iran was seen as a non-Arab, non-communist bastion around which a policy could be built offering the best chance of guaranteeing the security of oil supplies. So, feeling secure, the US State Department let its oil expertise run down. The office of Regional Petroleum Officer for the Middle East was abolished in 1962, leaving the Department's institutional concern in foreign oil questions almost exclusively in the Office of Fuels and Energy. This office was upgraded in importance in 1965, but still remained organisationally a sub-office within the Bureau of Economic and Business Affairs, reporting to the Secretary through the Deputy Assistant Secretary for International Resources and Food Policy, the Assistant Secretary for Economic and Business Affairs and the Under-Secretary of State for Economic Affairs. In January 1968, the Department's oil expertise was even further decreased when, as an economy measure, the post of Petroleum Attaché in the various producing countries was also abolished (Church Report, 1975, p. 16).

THE FRENCH

The French attitude was in great contrast to the complacency of the Anglo-Saxons during this period. Both national pride and commercial self-interest had been damaged by the scrapping of the Red Line Agreement in 1948 and the subsequent major expansion of American involvement in Middle Eastern oil in circumstances which were particularly humiliating to the French, who then started a massive search for oil in French-controlled territories. CFP was busy developing its share of IPC and repairing the damage the war had wreaked on its European operations. There was a feeling that CFP had 'sold out' and become indistinguishable in motive and style from the Anglo-

Saxon majors. So, given the complicated factional politics of the time, it was almost inevitable that another, more pliant national champion should emerge. This took the form of the Bureau de Recherche des Pétroles (BRP), which was the brainchild of the Gaullist Directeur des Carburants, Pierre Guillaumat. The French political culture has close personal links between top civil servants, politicians and businessmen and there is considerable interchange between government and industry via the practice known as 'la pantoufle' whereby top civil servants move into an industry some ten or fifteen years after they enter the service. Pierre Guillaumat is a particularly striking example of this. Directeur des Carburants from 1944 to 1951, he was also President of the Board of Directors of BRP, created in 1945 to broaden the search for new sources of crude oil. As this company successfully explored for oil in Algeria, then still a French province, Guillaumat moved into the atomic energy field as director of the Commissariat à l'Energie Atomique during the 1950s. He became Minister for the Army from 1958 to 1960, during the climactic years of the Franco-Algerian struggle, in which the discovery of significant oilfields from 1956 onwards was one factor in strengthening France's determination to fight against the Algerian demands for independence. After a spell as Minister in Charge of Nuclear Affairs, he took up posts within the oil industry again and seems to have been one of the guiding lights in the decision to create yet another new national champion by the merger of a number of entities, including BRP, resulting in the setting-up of L'Entreprise de Recherches et d'Activités Petrolières (Elf-ERAP) in 1966.

In fact BRP had concentrated initially on Morocco, Gabon and the Congo, but its greatest success was the discovery of the first major field in Algeria in 1956. By 1959 Algerian oil was being exported and, by 1964, had come to contribute 35 per cent of the French market, thus giving new impetus to *dirigisme* in oil affairs. Refiners operating in France not only had to transport at least two-thirds of their supplies in French-flag ships, but there was also a 'national obligation' that they take a certain proportion of oil from franc-zone sources. This meant that the majors who had been allowed into the French market had severe limitations put on their freedom to supply their refineries from sources that best suited them (Hartshorn, 1967, pp. 262–6; Chevalier, 1975, pp. 22–3).

But although the success of the search for franc-zone crude oil meant the French could reduce their dependence on the Anglo-Saxon majors, it also added considerably to the complexity of French diplomacy, since the chief finds were in a colony whose struggle for independence was marked by exceptional violence and brutality. When finally De Gaulle and the Algerians moved closer to agreement about independence, oil rose to the forefront of French thinking. Paris argued that the Saharan provinces where the oil was should be subject to a special status, and it was only four months before the Evian Agreements were signed (in March 1962) that the French finally conceded that these *départements* should belong to the newly independent nation. Even then, Paris took care to insist that Algeria should acknowledge that oil production was to continue under the existing Saharan oil code, which had been amended in favour of oil companies in the weeks preceding. This modified oil code was a bone of contention between Paris and Algiers for the next ten years (Chevalier, 1975, p. 731). It had been imposed on the Algerians at a time when they were really only concerned with establishing that the oil was theirs and had not yet considered terms of extraction. Paris had specified that French companies should be given preference in the assignment of concessions, that payment be made in francs, and that the conditions for price setting were such as to leave the Algerian tax authorities in an exceedingly weak position. The terms were amended in the Algerians' favour by the 1965 Franco-Algerian Agreements which increased the tax rate, revised some of the most objectionable financial provisions, gave the Algerian government first rights to the disposal of natural gas, created a fifty-fifty venture between the Algerian state oil company, Sonatrach, and an ERAP subsidiary, and created a Franco-Algerian body for implementing various industrial projects. By this deal, the French again ensured that they had access to good quality, relatively cheap oil for which they could pay in francs, while simultaneously ensuring that much of Algerian industrialisation would take place within a French framework (Hartshorn, 1967, pp. 264–5; Chevalier, 1975, p. 76).

As mentioned above, the French government consolidated the smaller French oil interests into Elf-ERAP, in 1966, the first totally government-owned petroleum company in France. With the advantage of Gaullist support and the preferential treatment given to its Algerian oil, the new company started well. Soon it was not content with its North African activities and sought to challenge the majors, including CFP, in their Middle Eastern preserve, by entering a service contract with Iran, the first in which the outside company had no ownership rights. This pioneering deal, like that of the ENI joint venture with NIOC in 1957, encouraged host governments to take a more aggressive attitude towards traditional concessions (FEA, 1975, p. 48; Fesharaki, 1976, pp. 70–82).

OTHER INDUSTRIALISED NON-PARENT GOVERNMENTS

Attempts to challenge the hegemony of the majors were not confined to France, for Italy, Germany and Japan all tried in their respective ways to resist the Anglo-Americans after the war. The period was one of aggressive expansion by the majors into the oil industries of the defeated nations at a time when potential indigenous competitors were unable to offer resistance. Middle Eastern oil was starting to flow in significant quantities; the economics of the industry were starting to favour refineries built near to ultimate markets and there was a vacuum in the defeated countries resulting from the demands of reconstruction on local policy makers. To some extent the majors were guided by occupation authorities. Admittedly, personnel involved in these later stressed that they tried hard to ensure that the majors were not given a *carte blanche* to pre-empt all competition, but there can be little doubt that the oil industry was close to these authorities at the time the political decisions were being taken to allow the defeated nations to re-industrialise. General MacArthur in Japan, for instance, established a Petroleum Advisory Group, composed of oil men on temporary leave from their companies. Paul Frankel has asked of their opposite numbers in Europe: 'can the leopard change its spots even if they are temporarily covered by a uniform?'[8] (1966, p. 10).

As a result of this interaction of military authorities, administrators of the European Recovery Programme and eager oil company executives, the independent European refiners which had appeared in the late 1930s were snuffed out. In Italy, state-owned entities like AGIP and ANIC who were involved in oil marketing, were tied to majors by long-term contracts (Frankel, 1966, pp. 60–1). In Germany, where coal was still the prime energy concern, the majors were easily able to re-establish their prewar position in the domestic oil market and thus ride on the back of the 'economic miracle' of the 1950s and 1960s. In prewar Japan there had been only one refinery with foreign interests. The industry which was created from 1949 onwards was heavily tied to the majors (Gulf and BP excepted) and Tidewater, the Getty company. Somewhat reluctantly, the government agreed to a formula in which companies put up half the money needed to build refineries in return for the permanent right to provide these with crude oil. By law, they had to have a 50 per cent Japanese stake, but the fortunes of each of these projects rested very much on the goodwill of the majors (Odell, 1970, pp. 118–20; Wilkins, 1974, pp. 315–16; FEA, 1975, pp. 92–3; Tsurumi, 1975, pp. 114–15).

JAPAN

Of the three defeated nations, Japan made the earliest and most persistent attempts to reduce its dependence on the majors. During the 1950s, the main thrust of the Ministry of International Trade and Industry (MITI), which oversaw the industry, was to reduce the cost of imported oil through its power over foreign exchange payments.

By publicising the prices Japan was actually paying for imported oil, MITI embarrassed the majors who had to explain to governments elsewhere why they were apparently favouring the Japanese market. This kind of restriction on the majors' freedom to 'source' their refining joint ventures as they saw fit was stepped up during the 1960s, with MITI insisting that all refiners take a proportion of the relatively unattractive (but not overly large flows of) crude oil produced in the Neutral Zone by the Japanese exploration company, the Arabian Oil Company (AOC). In addition, importers were also pressured to take supplies from non-traditional sources like the Soviet Union, and the powers of MITI were such that no company could meet such requests with a straight refusal.

However, MITI's policy – though aimed at reducing the role of the majors – was somewhat muddled. It encouraged the development of purely Japanese refining operations. But in the early postwar years Japanese interests were deliberately kept fragmented, apparently because it was felt that manufacturing industry would best be served by maximising competition between oil marketers. This was, perhaps, reasonable enough, but it also meant that none of these smallish companies had the financial strength to even consider going out into the world to seek new sources of crude oil, and until that happened Japan's dependence on international oil companies would never be reduced. This policy was notably distinct from that of France and Italy, where governments supported national champions (CFP, ENI and the forerunners of ERAP) to go out and either explore for oil in their own right or buy into existing concessions.[9]

The basic contradictions in Japanese policy were by no means eradicated during the 1960s, though the authorities did tighten up their policy throughout the decade. The 1962 Petroleum Industry Law improved the 'authorities' regulatory powers, but it was not until 1966 that a formal decision was taken within the Energy Council's petroleum committee to bring as much oil as possible under Japanese control. The following year it was agreed that by 1985 some 30 per cent of the total demand for crude oil should be met by Japanese companies. The Japan Petroleum Development Corporation (JPDC) was created the same year to spearhead the necessary search – a distinct change in policy from the 1950s when AOC had been given minimal official assistance in its fairly successful search for Middle Eastern crude oil (Tsurumi, 1975, pp. 117, 127). A quasi-governmental body, the JPDC was designed to identify potential projects, put the necessary deals together, provide finance where necessary and then withdraw as operations got going. By the mid-1970s, its support had enabled Japanese companies to move into Abu Dhabi, Indonesia, Alaska and Canada. This was indeed a step forward, but Japan's underlying problems still loomed large at the end of the 1960s. There was not much sign that the government money which was being poured into oil exploration via the JPDC would produce companies able to rival the majors. Moreover, the fact that the JPDC had primarily a banking role did little to overcome the over-fragmentation of the industry. After 1965, MITI encouraged the alliance of various small companies into what was to become Kyodo Sekiyu, but the participants were reluctant to pool both their refining and marketing operations.

MITI's goal of creating a sizeable Japanese company, integrated from the refining to the marketing stages, proved elusive, and, in any case, this strategy begged the question of whether this limited amount of integration would really be enough to produce a successful competitor to the foreign giants. On the other hand, although Japan did not have a contingency stockpiling plan against oil embargoes, it had consciously initiated a policy of diversifying sources of supplies. In the 1960s this meant encouraging importers to buy Soviet oil, and, in the following decade, this policy was to mature, with increased (but still minor) imports from China and Indonesia and an on-off relationship with Russia's Siberian developments (Howell and Morrow, 1974, pp. 50–4).

ITALY

There was no government ministry playing a dominant role in Italy, but there was the state-owned oil company ENI, which, under Enrico Mattei's guidance during the 1950s, took the profits from Po Valley gas and created a company which was more than willing to challenge the 'Seven Sisters' across the board. Consumed with bitterness against the majors, by whom he felt betrayed as they had not allowed him into the Iranian Consortium, Mattei was also a skilful and powerful actor within the Italian political scene. The alliance between ENI and the government manifested itself in such forms as tax cuts on petrol at a time when ENI was interested in boosting sales of this product, and an intergovernmental agreement with the Soviet Union in 1963 in which Italian goods were bartered for cheap Russian oil, which ENI successfully marketed, much to the irritation of the established international companies (Hartshorn, 1967, p. 278; Odell, 1970, pp. 53–4).

It was almost as if Italian oil policies were personified in Mattei, and certainly there is no other figure within the oil industry during the 1950s or 1960s (with the possible exception of Pierre Guillaumat) who came anywhere near to having as much political influence. Before he died, there were signs of some kind of reconciliation between him and the majors he had vilified so strongly during the previous decade, and his ENI successors achieved a relationship on a more cordial note. Mattei's death left a vacuum in Italian oil politics which increased the appreciation of the political contributions which Esso Italiana had made to the Italian political system since at least the early 1950s (authorised political contributions were $760,000 in 1963, peaking at $5 million in 1968) (Church Report, 1976, pt 12, pp. 242, 287). The 1960s was a boom time for the oil industry anyway and ENI was busy developing the various international projects to which Mattei had committed it before his death. In these circumstances, politicians, faced with a relatively undemanding ENI, were unlikely to be too nationalistic towards oil companies which were providing the benefits of stable petrol prices as well as significant contributions to party coffers. However, later, when the Teheran-Tripoli Agreements forced the majors to apply for higher prices in the Italian markets, all the old Italian suspicions which Mattei had so sedulously fostered came to the fore once again. Additional majors like Mobil were brought into the web of political contributions through the Unione Petrolifera, but these tactics brought only short run gains and did not counter the renewed challenge of ENI, prime beneficiary should any foreign companies get squeezed out of the Italian markets by over-zealous application of price controls (Church Report, 1975, pt 12, pp. 4–58, 241–340).

WEST GERMANY

Developments in West Germany were considerably less colourful. Unlike the Japanese, the Germans chose to develop a highly open economy and, until the later 1960s, were not particularly worried about the nationality of the oil companies doing business in their territory. Unlike the situation in Italy, no indigenous company found oil or gas reserves of sufficient importance to give them, like Mattei's ENI, the opportunity of breaking into the majors' preserves. German policy was primarily guided during the 1950s and early 1960s by the desire to slow the advance in the use of oil at the expense of coal, but, beyond that, there was apparently little serious thinking about what overall oil regime should be adopted. As late as 1957, Germany was still very much a coal-based economy, with imported primary energy only contributing 6 per cent of the total overall demand. However, it was clear that the days of coal's pre-eminence were numbered. A range of protective devices (including a tax on the consumption of all petroleum products, a tariff on crude imports, a special tax on heating oil and laws to subsidise the use of coal in electric power stations) were ranged against the oil industry – to no avail. There was even an effort, during 1958–9, to bring the leading oil and coal companies into a cartel in an attempt to give the oil companies a more binding reason for slowing down oil's incursions

(Hartshorn, 1967, pp. 269–72; FEA, 1975, p. 58; Mendershausen, 1976, p. 24). Within this general framework, Germany encouraged an extremely competitive oil market, unlike France or Britain where political and institutional factors raised barriers to new entrants to the industry. The Germans accepted a certain amount of Russian oil, thus sharpening price competition, and, with relatively easy entry into the retail end of the industry, a band of cut-price entrepreneurs appeared, to play havoc with the declared downstream profitability of the integrated majors.

There were some reservations about allowing the international companies to snuff out all indigenous German companies. In the early 1960s, the Federal government was insisting that such entrepreneurs should at least hold 25 per cent of the market, and some encouragement was gained later in the decade from the fact that at least one of them, Gelsenberg, was involved in a successful Libyan operation with Mobil. However, worries were growing towards the end of the 1960s. For one thing, the 1967 Suez War had once again reminded the Germans of their dependence on imported oil and, for another, a couple of takeover bids by foreign companies, particularly that of France's CFP for Gelsenberg in 1969, stirred some of the underlying xenophobia found in such circumstances in even the most liberally run society. The nationalists pointed to the fact that the market share of indigenous companies had declined from 40 to 25 per cent from 1963 to 1970. As a result, the Federal government decided to subsidise a joint exploration company, Deminex, formed by a number of German independent companies, as a first step towards reducing dependence on the oil of foreign competitors (FEA, 1975, pp. 57–62; Mendershausen, 1976, pp. 23–30). The fact that the government was willing to help such an operation was an indication that the Germans too were slowly working themselves into a frame of mind in which they would try to create an integrated national champion of their own.

So, during a period of complacency on the part of the Anglo-Saxon parent governments, France, Germany, Italy and Japan – the four next biggest non-communist consumers of oil – moved haltingly in the direction of trying to reduce their dependence on the traditional international majors. All either had developed, or were in the process of developing, a countervailing force of their own. France and Italy already had substantive national champions in existence, and France took the step of building up a second one through the consolidation of various companies into Elf-ERAP. Germany and Japan were less well equipped, but official thinking in both countries was favouring a similar rationalisation of indigenous companies and the positive encouragement of a national effort at oil exploration abroad. The concomitant of this was that the authorities in three of these countries had a more or less defined conception of the market share they wanted for indigenous companies. In Germany this was around 25 per cent. The Japanese had specifically set goals for 1985 and the French both required franc-zone crude to be used by French refiners (whatever their wishes) and used a system of assigning market and refining allocations to ensure that the market share of CFP and Elf-ERAP should continue to increase. Only the Italians appeared in the late 1960s to have no such definite plans, but this seems to have reflected the relative political ineffectiveness of ENI in the decade following Mattei's death, and was to be remedied in the 1970s, bringing Italian policy closer to the others.

Equally worrying for the traditional companies was the taste shown by France and Italy for bilateral intergovernmental deals with host oil producers, in place of reliance on the commercial role of the majors. The Franco-Algerian relationship was perhaps tolerable, given the peculiarities of the management of the French oil market and the special ex-colonial relationship. On the other hand, the Soviet-Italian trade agreements of the early 1960s, in which Italian goods were bartered for Russian oil, were a distinct shock for the majors. The Japanese government had not taken any such initiatives, but, as the international oil market tightened during the early 1970s, it showed by its enthusiasm for such bilateral deals that it was well attuned to the philosophy of its opposite numbers in France and Italy. Only the Germans, with their belief in a free market, demonstrated a faith in the ability of the majors.

Although a close analysis of trends within the thinking of these important non-parent consumer governments shows that attitudes were slowly turning against the traditional companies, no clear picture emerges of a similar trend amongst other middle-ranking powers. Although the issue of foreign dominance of domestic industries was growing in Canada and Australia during the late 1960s, this had not yet spread into hard policy decisions in the oil sector. Belgium responded to the 1967 Suez crisis by trying to persuade the companies to diversify their sources of supply, but otherwise left them alone. Of the countries on the fringe of the OECD, Spain went furthest, creating the roughly 55 per cent state-owned Hispanoil in 1965, charging it with the task of searching for oil abroad and then, in 1968, passing legislation limiting foreign involvement in Spanish refineries to a maximum of 40 per cent (FEA, 1975, pp. 15, 166). None of these countries, however, was sufficiently central to the oil industry for their policies to make much impact. The most influential governments still remained those of the parent countries. But, even here, there were developments within the OECD area which ensured that, despite a general complacency, new developments in the oil industry were posing subtle new problems for the parent governments.

THE NORTH SEA

The catalyst was the discovery of gas on the margins of (and then ultimately in) the North Sea which, from the early 1960s, forced both the British and the Dutch to re-examine their relationship with the majors as far as exploration and production in national waters were concerned. At the same time, the growing potential of the North Sea during the 1960s meant that a further range of European governments from Sweden and Norway to Germany had to establish a system to regulate exploration. The key influence was Britain, which was not only the most influential of the European parent governments, but also controlled the largest and most promising section of this sea. However much the British might want to protect the interests of established majors elsewhere, in the North Sea they themselves discreetly discriminated against those majors seen to be contributing little to the British economy, and encouraged a number of smaller competitors of all nationalities to apply for exploration permits in order to ensure that the majors would not hold back development of the North Sea in the interests of their holdings elsewhere. This was, in fact, a policy of discreet nationalism. It was unlike the US arrangement, where offshore licences were assigned to whatever company put in the highest bid. It was a discretionary approach, in which BP was favoured, to some extent, as a national champion, but was not given the kind of monopoly which would leave the country at the mercy of a single company. This was, of course, inconsistent on the part of the British who were, on the one hand, resisting Iraq's attempts to reclaim the extensive concessionary areas controlled by IPC (in which BP was the leading force), while simultaneously taking great care in the North Sea not to make the same mistake that the Iraqis had made of granting too much territory to a single company.

This discretionary British approach influenced the Norwegians, Dutch and Germans when they came round to setting up North Sea regimes of their own. In fact, the Dutch had gone further in giving the state a direct interest in domestic gas production. There had been quite an intensive search for gas carried out by Nederlandse Aardolie Maatschappij (NAM) from the end of the Second World War, a joint venture between Jersey and Shell. Once the major discovery of gas was made in 1959 near Groningen, the concession terms were altered to give the Dutch State Mines (DSM) a 40 per cent stake in production. In 1963, the sales organisation for natural gas was altered by the creation of a mixed venture, Gasunie, in which DSM had 40 per cent, the Shell-Esso partnership 50 per cent, and the state a 10 per cent stake. The companies were understandably uneasy lest the example of direct state involvement in the production of gas should be taken as a precedent in the oil sector. In fact, during the 1965–7 period, the Dutch government did try to insist that the companies which discovered offshore oil must accept state participation of up to 50 per cent in its exploitation, but industry pressure and electoral change ensured that this particular idea was dropped (Turner, 1975, pp. 93–7). Again the traditional freedom of the majors was eroded, this time by Britain and the Netherlands. In its way, this was just as significant as the more obvious pressure on the companies exerted by France, Italy, Japan and Germany. It was not as though there were any pressing security issues which made the British and Dutch authorities take a more assertive line, and neither country tried to persuade the OECD to take greater precautions against growing host government power. Rather, they were motivated by a sense that what was good for the majors – even ones as closely identified with them as Shell and BP – was not necessarily the best for their national economies. The abortive attempt of the Dutch to inject a hefty state participation in offshore oil production was a clear indication that even the governments of countries such as these no longer saw private companies as adequate national champions. When the Norwegians set up Statoil in 1973, a precedent was created which was very difficult for the British to ignore and the Labour government which was returned to power in 1974 lost little time in creating a totally state-owned company, the British National Oil Corporation (BNOC).

The 1960s thus ended with significant parts of the industrialised world showing a considerable loss of confidence in the companies as adequate agents. The development of the North Sea meant that a northern tier of European countries came to re-examine their attitudes to the majors. The result was that when the companies came to face the strongest pressure they had yet met from the host oil-producing world, their position in the industrialised world had started to slip. Inevitably, this meant that they were not given the wholehearted support that they needed from consumer countries – and that their failures to withstand the pressures from the hosts would bring their role in the industrialised world even further into question.

NOTES

1 There is a great deal of documentary raw material in Church Report, 1974, pt 7 and Church Report, 1975, pt 8.

2 The history of BP in Iran has generally been told by detractors (Williamson, 1927, and Longhurst, 1959, being exceptions). It will be interesting to see if Ronald Ferrier's forthcoming company history will be able to make out a more positive case.

3 The term 'independents' is used extremely loosely and in America refers only to companies heavily dependent on outside sources for supplies of oil (i.e. a refinery operation which buys from one of the crude-surplus

majors). In the world industry, the term is used to describe the companies which swarmed into the world arena from the late 1940s looking for new sources of crude oil. Some were very successful, particularly in Libya. By 1972 US companies like Occidental, Conoco, Standard Oil of Indiana, Marathon and Arco (Atlantic Richfield) had established themselves alongside the majors and various state oil companies in the lists of the major producers of crude oil outside the US and Communist world (Jacoby, 1974, p. 211).

4 An executive of one of the US majors assures me that he could still, in the early 1970s, sense a lack of commitment on the part of his top executives to the Iranian oil industry compared with their enthusiasm for other parts of the Middle Eastern industry.

5 Up to then, the Saudis had received a flat 12 per cent royalty payment on oil produced by Aramco. The 'fifty-fifty' concept was to give such producer governments a half-share in oil profits.

6 The companies now published (or 'posted') crude oil prices. As the years went by, the posted price was divorced from market prices, becoming instead the benchmark by which the levels of corporate taxation were set.

7 The APQ was a system for deciding the Consortium's production level each year. The members could lift their fixed proportion of this. The controversy arose because it had become by the 1960s uneconomic for any company (however short of crude oil) to take more than its official allocation (Church Report, 1975, p. 110).

8 See Kindleberger (1969, pp. 38–41) for the argument that the occupying authorities did not abuse their positions. For an overall picture see Wilkins (1974, pp. 314–16). For a Japanese view, see Tsurumi (1975, pp. 114–15).

9 Tsurumi suggests that the fragmentation of the oil industry gave MITI an important source of patronage by allowing the ministry to find homes for superannuated bureaucrats (1975, p. 118).

4
Majors and Host Governments pre–1970

So far it has been implied that the non-industrialised hosts were relatively impotent in the face of the implicit alliance between majors and parent governments. Although there is much truth in this, it would be a mistake to stress host governments' lack of power or disguise the fact that resistance to the companies had steadily increased. The revolutionary events of the 1970s did not come out of a vacuum.

Some aspects of the company-host government relationship are well documented. The terms of the key concessions are now on public record, there is a relatively clear picture of the oligopolistic organisation of the international industry and most of the important statistics are available. The debate is now less with the facts and more with the 'fairness' of the terms which the majors conceded to their hosts at any given time. From a transnational relations point of view, the interest is focused on faı more than just the shift in the balance of economic advantage between the majors and hosts. How did the majors manage to hang on to their relatively privileged position for so long? What tactics were used by both sides? To what extent were the majors able to influence host societies in matters having little to do with the oil industry?

On these questions the literature is less reliable. Much of it is highly tendentious, being either overtly demonological (the majors are the fount of all evil) or hagiographic (whereby the benefits the companies brought such as relatively well-paid jobs, medical services, etc., are seen as sufficient reason not to ask any deeper questions about the companies' exact role in the local scene). It is difficult to tread carefully through this veritable minefield of conflicting opinions to evaluate critically the historical record and draw broad conclusions about the relationship between companies and their hosts.

The first point worth stressing is that from the moment the companies were awarded concessions they were on the defensive. As early as 1900 the original Standard Oil ran into bitter political resistance in Romania, then a major oil-producing power. During the 1910s the oil industry's assets in the major oil province in Russia were expropriated and they were legally challenged in Mexico. In the same decade the Peruvians began a feud with Jersey's IPC which was to drag on for the next fifty years. The 1920s brought the first Latin American state company in the form of Argentina's YPF, while in the Middle East the Persians under Reza Shah demanded that Anglo-Persian increase its tax payments and agree to a revision of the terms of its concession. The 1930s saw expropriations and the creation of state company competitors in Uruguay, Peru, Bolivia and, most importantly, Mexico; the death of Venezuela's President Gomez, who had been very generous to the industry; and the revision of Anglo-Persian's concession at the request of what was now the Iranian government. The 1940s saw the start of significant interaction between Venezuela and Middle Eastern oil countries and the spread of the fifty-fifty tax principle. The 1950s saw this principle consolidated by the acceptance by parent governments of foreign tax exemptions which facilitated the industry's ability to make heavy payments to producer governments. In addition, this decade saw the creation of a further three state companies in Latin America, the Iranian expropriation of Anglo-Iranian and the growth of pressures against the companies in Iraq after the overthrow of the pro-British regime. In 1960 all these various strands of resistance came together in the creation of OPEC which, even if not particularly successful during its first decade, was an indication that the balance of power in the oil industry was now moving firmly in the direction of host governments. From 1947 to 1970 when both the posted and typical sales price paid for Middle Eastern crude oil actually fell, in absolute terms, the host government 'take' per barrel in Saudi Arabia went up over five times (Adelman, 1972, pp. 134, 208; Exxon, 1976, p. 15).

The picture is one of an industry on the defensive although there have always been some countries whose governments have not tried to push the companies too hard. This was true, for instance, of two Venezuelan regimes, those of Juan Vincente Gomez (1908–35) and Perez Jimenez (1950–8) where personal enrichment was put well before the national welfare. The pro-British monarchy in Iraq rarely gave IPC much of a problem and King Idris's regime in Libya during the 1950s and 1960s was very friendly to the companies. But these were exceptions. In general, from the 1920s onwards, the companies had to win concessions from governments which had a fair idea of the potential fortunes at stake and were fully capable of haggling with various suitors. In the Middle East Britain had treaties with various sheikhdoms and emirates giving it veto powers over concession agreements proposed to them. In Kuwait, for instance, the sheikh was able to call for advice from the British government on how to improve company offers. However feudal the regimes of countries like Kuwait or Saudi Arabia might have seemed to contemporary observers, this did not mean that their leaders, Sheikh Ahmad and King Abdul-Aziz, were not capable of telling a good bargain from a bad one and of playing American and British companies against each other when that would improve terms.[1]

However, the critical fact is that the companies did not really lose control of their relationship with host governments until the 1970s, when the concessionary system finally came close to being swept away. The long preceding decade of the 1960s had seen only minimal improvement for the host governments in the terms under which the majors did business with them. Some analysts have thus argued that it has always been a struggle between unequals – that the giant majors used a combination of political clout and economic coercion to keep the host governments in a permanently dependent position – that, according to the title of a recent book, the majors have been *Making Democracy Safe for Oil* (Rand, 1975).

MAJOR SKULLDUGGERY: COUPS AND WARS

The most extreme case to be made against the majors is that they have functioned as sovereign powers with the ability to trigger wars, influence the fate of the combatants and make and break governments. These sort of charges do not stand up well to investigation, but there are a handful of cases which do deserve further examination because I am aware of arguments (sometimes only put to me verbally) that oil companies were involved which seem to have had an interest in certain political outcomes and which might easily have been up to no good. In this section, I am looking at the events surrounding the original 1911 Mexican Revolution, the 1932–8 Chaco War between Bolivia and Paraguay, the 1951–4 Iranian crisis, the final break between the USA and Castro's Cuba in 1960–1, the Nigerian Civil War, the struggle for Angolan independence and certain aspects of Bolivian politics since the mid-1960s. There is no space to cast a jaundiced eye over the 1919 Leguia coup in Peru, the 1948 overthrow of the Accion Democratica government in Venezuela and the 1958 CIA-backed attempt to overthrow Sukarno in Indonesia (Howell and Morrow, 1974, pp. 123–4). However, even if I did have the space to give these events adequate treatment, I doubt if I would need to reverse my conclusion on the evidence available that the majors have virtually never stooped to skullduggery at this sort of level. On occasions they have been inextricably involved in crises arising from non-oil origins. But these are not enough to begin establishing the extreme demonological case.

Mexican Revolution: 1911

The upheaval created by the 1911 Mexican Revolution has attracted considerable popular interest, with numerous books and films on romantic revolutionary heroes like Villa and Zapata. There have also been a number of claims that the battle of American and British oil interests was a key factor in the original overthrow of President Diaz by Madero, and in the latter's subsequent replacement by first Huerta and then Carranza. In its starkest form, the charge reads that the first and third of these were representatives of British Pearson/Cowdray interests, while the second and last were backed by the Americans Doheny and Standard Oil[2] (Denny, 1930, pp. 240–53; Calcott, 1968, chs 3, 4; Calvert, 1968; Nearing and Freeman, 1969, pp. 84–121).

Calvert has carried out a very thorough examination of these charges which, while demonstrating that an understanding of Mexican politics of the time requires knowledge of oil company rivalries, dismisses the wider charges as grossly oversimplified. For instance, though President Wilson authorised the State Department to write to Archbold of Standard Oil, referring to allegations that the company had financed Madero in his efforts to overthrow the pro-British Diaz, evidence suggests that the former had embezzled more than enough funds to have achieved this feat on his own – even if there was a Standard agent trying to make contact with him.[3] Calvert is also suspicious of the claims that these presidents each favoured one oil company, only to be overthrown by someone favouring the other. For instance, the leading British oil magnate in Mexico, Pearson, was favoured by Diaz and was willing to help the latter in his flight to exile. However, evidence shows that Pearson soon came to an understanding with Diaz's apparently pro-American successor, Madero, with whom he had good relations to the end. And although the Americans were convinced that Madero's successor, Huerta, was a simple tool of Pearson, the latter was in fact banned by Huerta from selling his railroad interests in Mexico and showed a notable lack of enthusiasm about subscribing to a loan of that administration (Calvert, 1968, pp. 109, 230–1). However, Calvert's study shows that foreign oil interests, if not the president makers of demonological mythology, were certainly important pressure groups in the political environment of those times.

Chaco War

The Chaco War has been described as an incident 'when Standard Oil and Royal Dutch Shell fought each other for oil deposits in that large area of sun-baked scrublands' (Hobsbawm, 1975, p. 23). This description shows how contemporary allegations of corporate skullduggery, however tenuously grounded in fact, have taken on a life of their own.

This is a war for which there is one quite obvious explanation. Bolivia had lost access to an ocean in the nineteenth-century War of the Pacific, and had then had its last hopes of retrieving the situation dashed when Peru and Chile did a deal in 1929 which excluded the Bolivians. Bolivia then naturally looked eastwards toward the Paraguay river and tried to take over the virtually uninhabited Chaco which lay in the way. Paraguay resisted and the two countries went to war from 1932 to 1938 (Wood, 1966, p. 19; Pendle, 1968, pp. 25–6). Oil was involved in this dispute because Jersey was producing a limited amount of crude oil in Bolivia and had made discoveries sufficiently near to the Chaco area to justify speculation that the oil province might extend into the Chaco. This was sufficient to trigger Paraguayan claims that Jersey was financing the Bolivian war effort, and these charges were echoed by Senator Huey Long in the US Senate, along with a claim that Jersey was trying to secure a port from which to export Bolivian crude oil. At the same time the Argentine press was insisting that this was a battle between Jersey and Shell for Chaco oil rights. These claims were reinforced by the fact that Shell's Deterding was somehow involved in bolstering the Bolivian credit rating when Bolivia sought to buy arms from Vickers Armstrong (Wood, 1966, pp. 20, 27, 65–7, 118; Fifer, 1970, pp. 210–22). They were echoed in statements such as that by J. W. Lindsay in *International Affairs:* 'Petrol is the invisible cause of the Chaco War' (vol. 14, 1935, p. 235).

These charges remain unproved. The State Department which, under the New Deal, was taking a relatively hard line against corporate depradation in Latin America, seems to have been unimpressed by the case against Jersey, and the latter's claim that Bolivian production was commercially insignificant has not been questioned. If Shell was so determined to secure the oil rights in the Chaco, why did it approach Bolivia, where Jersey was already installed, instead of Paraguay which had, by far, the better legal claim to the area? If Jersey did give Bolivia financial help, why did the Bolivians repay the company by expropriating its property in 1937? As in the case of the Nigerian Civil War, oil may well have played a role in exacerbating and prolonging the conflict as both sides became convinced that the Chaco did indeed contain significant oil riches, but it is difficult to prove that it caused the war.

Mussadiq's Overthrow

Scepticism is clearly invited by the events surrounding Mussadiq's overthrow, for here quite clearly was the successful CIA-backed coup against the man who had expropriated a major oil company, Anglo-Iranian. What could be clearer than that there must have been a direct link between the company's fate and the eventual fall of Mussadiq?

I have earlier (p. 45) given some idea of the complexity of the Anglo-American diplomacy surrounding this case and this should go a long way towards showing that straight corporate interests were not the only issues at stake. It is also possible to argue that Anglo-Iranian's relative impotence in the face of very wide-sweeping Iranian attack is shown by the fact that it was soon clear to the diplomats involved that there was virtually no chance of any Iranian regime which might replace Mussadiq accepting the company back to the position of dominance it had previously had in the country. Therefore, Western policy makers had to devise the consortium formula to make Anglo-Iranian's eventual re-entry into the Iranian economy more tolerable to that nation. Again, it is clear that the eventual coup was not one which was put together by the oil industry. The decision to throw the CIA into backing the assault on Mussadiq was one which was taken at the highest levels in Washington and London. Doubtless Anglo-Iranian and the other majors thoroughly approved this action, but by 1953 they were mere bystanders.

On the other hand, this does not mean the companies had no influence on post-nationalisation Iran. They may not have provided the money or material support which the coup needed, but they had wrecked the Iranian economy by the success of the industrywide boycott of Iranian oil exports. So when I argued earlier that the Anglo-American politicians felt they had to find a solution to the Iranian problem because they could not allow a country like Iran to 'go communist' in such a geographical position and at the height of the Cold War, I have to note that the chaos caused by the companies' embargo could only play into the hands of the Soviet Union. And this was then used as the reason for overthrowing Mussadiq's administration.

So a dispute which was about the legitimacy of one oil company in Iran escalated into an issue where it appeared that the Soviet Union might be gaining a victory at the expense of the 'Free World'. Given the times, the coup that followed this was virtually inevitable. Anglo-Iranian regained some of its stake in the country. Its American competitors got a substantial foothold which they had not really sought with any enthusiasm. And Kermit Roosevelt, the leading American connected with the coup, was subsequently to become a director in charge of Gulf's government relations (Church Report, 1974, *IPC*; Tanzer, 1969, p. 326; Sampson, 1975, p. 196).

Cuban Revolution

The relationship of Texaco, Shell and Jersey with Castro's Cuba deserves closer attention because it was Cuba's decision to expropriate their Cuban operations (minimal oil production was involved) which marked one of the main turning points in the relationship between the USA and Cuba. This is a difficult period about which to write since so much of what went on in Washington was at the clandestine level, and it may well be that opening the oil industry files would reverse some of my judgements. On the available evidence oil interests played little or no role in the first stages of the deterioration of the Cuban-US relationship, but provided the catalytic event which ultimately triggered off the final break between the two countries.

It may well be that such a rupture was inevitable once Castro had seized power in January 1959, although it was to take two years to come about. It is extremely doubtful that oil interests played any significant role in the initial deterioration of relations during 1959 which was the year which set the tone for all that followed. By the summer of 1959 the US intelligence community, and Vice-President Nixon, were convinced that Castro was pro-communist at the head of a regime with at least some formal Communist Party members. His agrarian reform antagonised the long-established American sugar interests. Executions of opponents were arousing public unease among US citizens who otherwise might have had an open mind toward Castro. By the autumn of 1959 his opponents had started bombing raids.

It was not until November 1959 that there was specific action against the oil industry in Cuba in the form of a law which insisted that the companies drill on their concessions; but this was one of a variety of measures which offended the whole international business community. The hotel companies, the cattle companies, Bethlehem Steel and International Harvester had far greater cause to complain.

Oil investment was relatively unimportant, particularly compared with sugar which had attracted a third of all capital invested in Cuba. There was a growing interest in oil exploration after a discovery in 1954 (the first strike in Cuba had been made in 1914), but prospects were unclear and most Cuban oil needs were met from imports. Given the majors' enormous reserves in the Middle East, it is hard to believe that Cuban operations were particularly significant to them.

In his massive book, Hugh Thomas argues that the USA finally lost all goodwill in Cuba by February 1960, when the last US employee of the phone company was expelled. Cuban rhetoric was still very much concerned with sugar interests and the US sugar lobby began to campaign against renewing Cuba's sugar quotas. Even if oil interests were not yet active, the critical point had probably come in January 1960 when Cuba signed an economic agreement with the Soviet Union which called for imports of Soviet crude oil in return for sugar exports. Some of this crude oil started arriving in April, but it was not until 23 May that Texaco, Shell and Jersey were told that a large consignment of Russian oil was on its way and that they would henceforth be expected to process oil from this source in their refineries. The Cubans hoped to supply 20 per cent of their market in this way. The companies hesitated and then refused to process this oil. In his memoirs, Philip Bonsal, US ambassador to Cuba, claims that the companies were resigned to carrying out Cuban instructions under protest while seeking redress through the courts. However, Bonsal was informed by an oil executive that representatives of the two US companies were summoned to the office of Robert Anderson, Secretary of the Treasury, and told that the US government would be happy to see them refuse to handle Soviet crude oil and that Shell was being requested in London to take the same line. The relevant area officials within the State Department appear to have had this policy sprung on them without advance notice (Thomas, 1971, pp. 1158–1298; Bonsal, 1971, pp. 145–50).

From this point, US-Cuban relations went downhill precipitously. While the Cubans pondered their response, Congress, with the State Department's approval, passed a Bill permitting the president to cut or eliminate Cuban sugar quotas for 1960. Castro talked about the declaration of economic war and the majors' refineries were taken over at the end of June. By mid-July, Eisenhower had wiped out Cuba's sugar quota for the rest of the year. In August, Cuba formally expropriated the refineries along with the electric and phone companies and the sugar mills. A steady deterioration of relations then occurred, culminating in their formal breaking in January 1961 (Thomas, 1971, pp. 1284–91).

On balance, the oil companies appear to have been relatively passive actors in this drama. Ambassador Bonsal, an opponent of the Washington hard-liners, gives the impression of an industry trying quite hard to placate the revolutionary leaders. The companies had made advance payments to the Cuban Treasury in January 1959 to relieve its cash shortages. They did not press for the dollars they would normally have required to import crude oil and by May 1960 the Cuban government owed them some 50 million dollars. Bonsal certainly suggests that their initial reaction to Cuba's insistence that they take Soviet oil was relatively subdued. Although the issue of Soviet oil was one which did worry the oil industry in the early 1960s it would appear that the key US decision makers, by June 1960, were looking for an excuse to bring US-Cuban differences to a head, and the ultimatum to the refiners made as good a sticking point as any. Thus the majors were made agents of US policy. There is no evidence that they played any significant part in making it.

Nigerian Civil War

The particularly vicious Nigerian civil war started in 1967 over a dispute about the exact destination of a royalty payment from a

Shell-BP joint venture. Shell-BP initially discovered oil in 1956 in what was then known as the eastern region, dominated, though by no means exclusively populated, by the Ibo tribe. By 1966, three companies were producing oil commercially: Shell-BP had by far the largest operation producing 485,000 b/d in April 1967, of which 340,000 b/d came from the eastern region. The French company SAFRAP (a subsidiary of Elf-ERAP) was producing 41,000 b/d entirely from the east, while Gulf had one field producing 56,000 b/d in the midwestern region (which remained on the Federal side in the Civil War) (Pearson, 1970, pp. 15, 70).

With or without the presence of oil, the Nigerian Federation was crumbling as coup, counter-coup and inter-tribal massacres led to a polarisation of the eastern region, primarily Ibo, and the Federal government, primarily run by Hausas from the north. By the beginning of 1967, the east was demanding a looser federation, and focused its attention on the destination of Shell-BP's oil royalties, up until then paid to the central government, which distributed shares to the regions. The east insisted that Shell-BP should pay it directly its share of royalties. The Federal government saw this as a threat to its sovereignty and warned that it would invade the east should Shell-BP fail to pay it the complete sum of royalties as usual. The companies were caught in the middle and tried to buy time. As the pressures mounted toward the summer of 1967, Shell-BP wrote to Colonel Ojukwu, the eastern leader, offering a down payment of £250,000 with £7 million to follow – roughly what the east would have received under the normal Federal government formula. Both sides were displeased. Ojukwu found this token payment insulting, while the Federal government believed that payment had actually been made and therefore imposed a naval blockade on tankers seeking to use the terminal in eastern Biafran territory. The Civil War started within days.

The diplomacy surrounding this war was extremely complex. The two parent governments of Shell, dominant partner in the joint venture with BP, took opposing lines, with the British backing the Federal government to the hilt, while the Dutch (not the world's most significant arms traders) instituted an arms embargo from January 1968, apparently on the humanitarian grounds that this would leave the federal authorities in charge but without the resources to reduce Biafra to complete submission (Cronje, 1972, p. 144).

It would appear that Shell-BP took a passive role, relying on legal precedents and on guidance from the British government. There seems to be no evidence that the company encouraged the Biafran secession in any way. Once the breakaway had occurred, it chose to follow the legal precedent from events after the 1949 communist victory in China which counselled that companies should make their peace with *de facto* rulers, even if parent governments had no dealings with the new regimes. At the same time, however, the company got ambivalent advice from the British government, whose dilemma was increased by the closure of the Suez Canal in June 1967. As a result of this crisis, the government decided that since safeguarding oil supplies was essential, if Ojukwu won *de facto* control, the companies should then be able to pay royalties to the Biafra regime. It is claimed that the Secretary of State for Commonwealth Affairs verbally instructed Shell-BP to open a suspense account, into which the disputed revenue should be entered, and that a junior minister reluctantly agreed that the company should make a token payment to Ojukwu (Cronje, 1972, p. 24; de St Jorre, 1972, p. 140). But British policy changed as the Foreign Office, then distinct from the Commonwealth Office, took control. Shell-BP applied to the Bank of England for exchange control permission to export the necessary currency to make a royalty payment to Biafra and was refused, apparently on instructions of the government. At this point, fighting had broken out. From then on, official British policy was firmly behind the Federal forces and all the indications are that the Shell-BP partnership followed this line, although this decision caused considerable problems. The bulk of Shell-BP installations were in Biafran territory, and their local manager was held as a hostage, only being released after Frank McFadzean, then a group managing director of Shell, crossed into Biafra and came out with his colleagues (Turner, 1973, pp. 31–6).

This incident illustrates the vulnerability of oilfields, refineries and terminals in a politically disturbed environment, for the final lesson from the Nigerian Civil War is that the Shell-BP joint venture, despite the size of the partners, was politically impotent. Certainly the amount of oil produced meant that the distribution of royalties and taxes so generated was a prize worth fighting over, but war would have broken out regardless of Shell-BP's strategy, or regardless of whether the operations had been developed by some other company. All Shell-BP could influence was the timing of the war's start. The existence of oil was a crucial factor in exacerbating existing tensions and reinforcing British and French government concern. The identity of the oil companies was irrelevant.

Angola, Bolivia and Gulf

In a different way, there was a similar amount of controversy about the role of Gulf in Angola during the years leading up to Angolan independence in 1975, and in Bolivia in the years leading up to 1970. In 1967 Gulf discovered commercial quantities of oil in the Cabindan enclave of Angola and started to export it the following year. Angola was then a Portuguese 'overseas province' and, as one of the last European colonies in Africa, a controversial area. A variety of anti-Portuguese groups attacked Gulf for prolonging colonial rule through its royalty and tax payments to an administration which, partly because of the relative poverty of metropolitan Portugal, needed money badly.

In the aftermath of 1974's 'Revolution of the Carnations', a transitional government set up in Angola became the target of a struggle between three warring factions: UNITA, FNLA and MPLA (there was also a secessionist movement in Cabinda called FLEC). By the end of 1975, the MPLA had emerged as being in *de facto* control of the Angolan central administration, but as it espoused Marxism and received military aid from Cuba, it was not immediately accepted by the USA which was aiding the other two main groups. Gulf was therefore in a quandary. Cabinda escaped most of the violence seen elsewhere in Angola, but, through the latter part of 1975, MPLA sympathisers had gradually become dominant in the local Cabindan administration. Gulf seems to have accepted the inevitability of this and oil continued to be pumped as the rival factions battled for control of the country as a whole. In particular, Gulf duly switched its payments from the Portuguese provincial authorities to the transitional government whose finance ministry was in MPLA hands. Regular payments were made until September 1975.

Angolan independence came that November, and the MPLA took some three or four months to assert its military superiority over its rivals. During this period the question of Gulf's royalty and tax payments came up (they represented about half the total income of the Angolan government before the Civil War). Gulf had been uneasy about the aid the US government was giving UNITA and the FNLA forces, and by the end of 1975, there were frequent discussions with the State Department about Gulf's commitments. The State Department had suggested that payments should not be made to the new government, a suggestion which Gulf resisted on the grounds that such an action would endanger the safety of its personnel and installations. In December 1975, all three warring factions demanded that month's payment. Gulf withdrew its personnel, temporarily suspended operations and placed all royalties and taxes in a special interest-bearing escrow account until a generally recognised government emerged. While the State Department requested that personnel be withdrawn for reasons of safety, Gulf has made it plain that the decision about payments was solely the responsibility of the company. Both decisions were necessary. The situation in Cabinda had deteriorated, and insurance companies were balking about covering the tankers and other equipment involved. There was a possibility that Zaïre would block passage into and out of Cabinda which would have increased the risks of personnel there. The fact that the State Department also had a vested interest in demonstrating to the outside world that the MPLA were not in control of the country was coincidental to the real physical dangers of the situation at that time.

By February 1976, conditions had changed and the State Department suggested that Gulf should release the blocked funds and negotiate directly with the MPLA regime. On 10 March, the payments were made. The new regime assured Gulf that it had no intention of nationalising the oilfields, and pumping resumed. The fact that the MPLA reacted thus calmly, despite the undoubted hostility of the US administration, would seem to indicate that Gulf had been trying to follow a neutral path in making a relatively smooth transition from dealing with the former Portuguese colonial administration to working with its Marxist successor.

However, if the evidence suggests that Gulf's payments in Angola were above board, the same cannot be said of its activities elsewhere in the world during the 1970s and late 1960s, for it was then that Gulf became embroiled with dubious payments in Bolivia, Italy, South Korea and the USA. The Bolivian case suggests that further study would be worthwhile of the five eventful years from the discovery in 1966 by Gulf, then producing oil, of a quantity of natural gas in excess of Bolivian needs. The company leased for former President General Rene Barrientos Ortuno, an Air Corps general who had been trained in the United States and who was running for re-election, a helicopter which he wanted for purposes of campaigning among the Indians in the mountainous Altiplano region of Bolivia. It was alleged that Gulf also accepted a commitment to underwrite a payment in excess of 1 million dollars to Barrientos, ostensibly part of a transaction involving the construction of a gas pipeline from the district of Santa Cruz to a point on the Argentine border near Yacuiba for purposes of delivering gas to Argentina.

Barrientos was killed in a helicopter crash in April 1969. A subsequent regime under General Alfredo Ovando expropriated Gulf's properties in October of the same year. Months of hard bargaining followed before Gulf obtained a compensation arrangement in 1970 which was acceptable to Gulf but did not result in restoration of its properties. In 1971, a series of coups took place, bringing in a right-wing military officer, General Hugo Banzer, who publicly announced that the nationalisation of the Gulf properties had been a mistake. However, Gulf's properties were still not returned.

I have argued elsewhere that there is no case for suggesting that Gulf was involved in any of these changes of regime. One of the deposed presidents, General Alfredo Ovando, whose government accomplished the expropriation of Gulf's property, claimed that Gulf was implicated in his overthrow; however, there is no evidence to support this. On the contrary, the Torres government, which overthrew General Ovando, was even more leftist in its character. Furthermore, by 1970–1, there were far weightier bodies than Gulf interested in the political fate of Bolivia – such as the neighbouring governments of

Brazil and Argentina. What is clear, though, is that Gulf was perhaps unwise in its political relationships here as elsewhere. The only known intended recipient, General Barrientos, was killed before any such payment could have been made to him. But the impression remains that Gulf was not the maker and breaker of Bolivian presidents, in any sense of the word (Turner, 1973, pp. 36–7).

PARENTAL PROTECTION: GUNBOATS, MARINES AND EMBARGOES

Given the symbiotic relationship between companies and their parent governments, it has not been unusual for the latter to come to the former's aid with resources befitting a sovereign state. The use of armed forces has been rare. There is no case where the US marines were sent in specifically to protect US oil interests, though it would be possible to argue that US intervention in the early days of the Mexican Revolution, such as the Vera Cruz incident and Pershing's expedition against Villa, were at least partially motivated by American concern about the way the revolution was turning against the oil industry in a country then in the midst of an oil boom.

The British have been less restrained. Troops from India were used in Persia to guard the original explorers for oil and to put down a strike in 1922. They were sent ashore in a muddled invasion of Abadan in 1941[4] and were stationed in Basrah in 1946 when a labour dispute in Iran threatened to get out of hand. In the background were the seemingly ubiquitous gunboats, of which one had stuck on a Persian mudbank as early as 1905 in the interests of impressing local tribesmen with respect for British oil prospectors. Gunboats showed the flag off Mexico in 1924 to protect a refinery and off Iran in 1932 during the re-negotiation of Anglo-Persian's concession, again in 1946 and yet again in 1951 at the height of the nationalisation crisis (Elwell-Sutton, 1955, pp. 76, 148, 210; Cable, 1971, p. 182).

However, the fact that there have been so few incidents like the above can be used to argue that such involvement by parent governments was extremely impracticable. Host countries were sovereign states, and it was only in the case of a very weak government, such as in Persia in 1922, that any imperial power could hope to bring in troops without declaring war on the host country. And in the case of Iran, the notably bad relations of Anglo-Iranian and its Iranian hosts in the late 1940s were at least partly a function of how openly the British government appeared willing to rattle the sabre in the maintenance of the company's interests.

Although parent governments have generally ruled out armed force as an efficient way of protecting the interest of their companies, they have had a number of non-military options open to them. The

British broke diplomatic relations with Mexico after Shell's expropriation in 1938, though there is no evidence to suggest that this brought either Britain or Shell any particular advantage. The British foreign office also pleaded Anglo-Persian's case in 1932, when the dispute over revisions of the company's concession reached the League of Nations. In general, a parent government tended to side with the companies in disputes with hosts, although there have been cases to the contrary. Roosevelt's decision to put his Good Neighbor policy ahead of the commercial interests of the US oil industry in Mexico, thus limiting the support the State Department was willing to give the relevant American companies, is a case in point. A new form of leverage has appeared with the rise of international aid programmes. The USA especially has threatened to withhold aid if the US oil industry is treated badly. But the 1962 Hickenlooper Amendment, which made aid suspension automatic wherever a country expropriated US property without prompt and adequate compensation, was a two-edged sword. US-Peruvian relations were bedevilled throughout the 1960s as Jersey's IPC battled against expropriation. The prospect of withdrawal of aid undoubtedly made the Peruvians more cautious, but it also meant that the USA was forced to rattle the financial sabre at a time when it wanted friends in Latin America.[5] However, aid programmes have not always worked in the companies' favour. One of the earlier exposures of oligopolistic pricing policies by the majors sprang from the Marshall Plan. The fact that the ECA was funding European purchases of oil made prices paid a matter of public concern, and the resultant scrutiny led to price cuts on the parts of the companies (Tanzer, 1969, pp. 399–400).

One of the most interesting areas of parent government leverage after the Second World War was through the World Bank which, as the leading international source of funding for Third World development, was opposed until the 1970s to lending money to LDCs (less developed countries) for oil exploration and processing projects. The Mexicans claimed that demands for a development loan for Pemex were turned down for ideological reasons and the Bank put pressure on India to stop it going in for overseas oil exploration in the early 1960s (Tanzer, 1969, pp. 90–5). Undoubtedly much of the World Bank opposition to these projects stemmed from an ideological position against state oil companies, as this organisation had been moulded by men like Eugene Black who were outspoken supporters of private enterprise.

Despite these advantages, parent government pressures have been of limited importance. The extensive diplomatic attention which the USA paid to Mexico from the 1910s onwards failed to avert the 1938 expropriation. All the military and diplomatic pressure that the British brought to bear on Iran over the years failed to ensure that, even after Mussadiq's overthrow, the Iranian authorities would allow Anglo-Iranian more than a minority stake in an industry it had hitherto dominated. Similarly, despite the Hickenlooper Amendment and continuous American manipulation of aid policies during the 1960s, the Peruvians still went ahead and expropriated IPC. There was nothing, moreover, that the US government could do to save the majors' refineries in Cuba.

It can be argued that I have underestimated the symbolic importance of the use of a gunboat, coup or cessation of aid and that the reason why there have been relatively few instances of such extreme measures is that the parent governments have been so sufficiently in charge that the need for such measures has not arisen. Nothing happened for many years (it may be argued) simply because host authorities were in such a state of dependence that they did not conceive of challenging the parent governments.

This argument should be taken seriously. There was undoubtedly symbolic importance in the 1953 coup in Iran which obviously worried regimes throughout the Middle East, particularly in the small states on the Gulf which were well aware of the presence of the British Navy. Traditionalist regimes such as those in Saudi Arabia and Libya rarely questioned their dependence on the West until the late 1960s. On the other hand, the post-1958 regimes in Iraq were able to challenge the oil industry and survive, and Venezuela worried far more about being excluded from the US oil market than about marines storming ashore. By the mid-1960s the Shah of Iran had come to realise that the West was tied to *his* fortunes and that he could pressure the Western oil industry with growing impunity.

However, the biggest single argument to back my view that gunboat/coup strategies were seen as fundamentally impracticable by parent governments is that there has been since the late 1960s a revolution in the oil industry with virtually no recourse by parental governments to such practices. If the arguments described above were correct, then host government pressures on the companies during the 1970s would have been met by a wave of British, American, Dutch and French military interventions or attempted coups. In fact there were no such developments despite the fact that the events of the early 1970s which culminated in the Arab oil embargo clearly resulted from the actions of governments which saw themselves as anything but dependent. There was talk in 1973–4 of sending in US troops to take over the Saudi oilfields but this was never seriously considered by officials who knew that it was a strategy which could not be used. If the contemporary Saudi regime toppled, any replacement was bound to be less pro-Western and the rage in the rest of the oil-producing world would have triggered off a series of unforeseeable diplomatic side-effects. In other words, direct military intervention was not then a serious option for parent governments faced with the challenge from the producing world.

But it has rarely been different in the oil industry. There have only been two 'no-holds-barred' interventions in the domestic politics of a host country – the invasion of Iran during the Second World War and the 1953 Iranian coup. These were 'extraordinary' events in the full sense of that word, one triggered by the exceptional circumstances of a world war and the other by strategic considerations at the height of the Cold War. These two cases must be compared to events such as the Mexican expropriations of 1938 or the Iraqi takeover of undeveloped concessionary territory in 1961. Both these were challenges to the established industry as severe as that posed by Iran, but they were posed by countries and at times which made the military or secret-service option impossible for parent governments to use. The fact that parent governments reacted so tamely to the host government challenge of the 1970s really shows just how extraordinary the Iranian cases have been. In general, military intervention or substantial subversion has been impracticable for parent governments to use in protecting their oil interests. Although it is not easily demonstrable, host governments became aware of this as they started to challenge the established oil industry. The really interesting question is why they did not seek to raise this challenge earlier.

ELITE MANIPULATION: BRIBERY, CORRUPTION

The biggest gap in the historical record is over the extent to which companies could control the domestic politics of host governments by means falling short of the spectacular practices discussed above. To what extent have such extreme measures been unnecessary because there have been subtler methods of control, relying on the identification and control of key decision makers and political elites? Western historians have generally either not been interested in asking such questions or have been hampered by the problems of inquiring into the politics of very alien societies, and into subject matters which the companies would want to hide even more thoroughly than they did the details of concession agreements and intra-Consortium regulations. However, there are now some studies which are sufficiently detailed to throw light on these types of vexed question (Pearson, 1970; Pearton, 1971; Pinelo, 1973; Goodsell, 1974; Tugwell, 1975).

At one end of the spectrum of published information are the claims of the companies. BP executives insist that company policy has always specifically warned executives against political involvement in host countries, and most of the other majors strive to give the same impression. A good statement of the official 'line' comes from an industry-oriented study of Creole Petroleum Corporation, Jersey's subsidiary in Venezuela:

> However, in one important area Creole does not act like a citizen at all. This is in the field of Venezuelan politics, which Creole and its officials avoid completely . . . Political neutrality has long been recognised as the only sensible attitude for a United States company to have when it operates abroad, especially in Latin America . . . Although [Creole] identifies its interests with those of Venezuela, it does not claim the right to advise Venezuelans about their form of government or about the men who are to run it . . . For Creole, Venezuelan politics is like Venezuela's climate, terrain, and geography – something which the company must take as it finds if it wants to produce oil successfully (Taylor and Lindeman, 1955, pp. 22–3).

This policy statement may well be quite true, but it is worth remembering that it was under some State Department pressure that Jersey's representative in Caracas was sacked or resigned in 1942 because he was unwilling to come to terms with Venezuela's post-Gomez regimes (Wilkins, 1974, p. 270). It is well to remember also the disclosure of Exxon's long-standing involvement in political payments to Italian political parties. There is substantial evidence in at least this latter country of the oil company not espousing political neutrality but rather of giving the bulk of its backing to parties of an anti-communist persuasion.

Analyses of corruption are necessarily difficult since there are so many different situations and different ways of viewing payments. For instance, it has been the case in a number of host countries, particularly early in the century, that the head of state ran the country as his personal fiefdom. In the case of Saudi Arabia during the 1930s and 1940s, what difference did it make if payments were made as royalties rather than as political setttlements made directly to members of the royal family? The destination of the money was not that different (Taylor and Lindeman, 1955, p. 25).

What sort of sums were involved? A negotiator for one of the companies involved in the Middle East in the 1930s claims that although he was given *carte blanche* by his company to make political payments, at the most this meant putting a ruler on an annuity while he pondered the terms which the company was offering. In general there would be routine offerings of gifts such as clocks and a certain amount of money would be spread around the courtiers. The gifts may not have meant much to the companies involved but were calculated to impress the citizens of the relatively impoverished host countries. Such gift-giving seems to have been fairly ritualised and became part of ordinary negotiations within the Middle East. In fact, of course, presentation gifts are still very much a part of the ritual of diplomacy between heads of state, and not only in the Third World. It is probably safe to assume that all companies would have been expected to make ritual payments of this type as a matter of course, but that such sums would rarely have tipped contracts one way or another.

There were, of course, situations when payments were competitive as, apparently, in the 1966 round of bidding for concessions in Libya (First, 1974, pp. 194–7; Church Report, 1975, p. 99). But large-scale payments are a sign that established companies are vulnerable to challenge from new competitors. What the former want are regimes which can be trusted to stand by the terms of contractual arrangements. The last thing they want is to work under governments whose decisions reflect not the rule of law but rather the influence of persons who can offer the largest sums of money. It may be worthwhile for an established company to ensure that a given regime is kept in power, but there is a narrow line to be drawn between the occasional advance payment of taxes or the ritual goodwill payment which may help this and a situation in which a regime routinely comes to demand payments to influence its decisions.

Moreover, the practices of the majors were changed as even the most traditional host countries were forced to modernise. The concept of the State hardened, giving clearer definition to the bounds of propriety surrounding public office. For commercial reasons, the companies could no longer turn a blind eye to the distinction made by host regimes between public and private benefit. Identification with the corrupt practices associated with particular regimes became dangerous, lest these regimes should be toppled. The companies had been taught this lesson in Venezuela by the reaction to Gomez's death which was marked by riots against foreigners in general and the oil companies in particular, and led to a decade of political debate culminating in tightened concession and operating terms, and, finally, the fifty-fifty principle which, once accepted in Venezuela, was to sweep the rest of the oil world.

Political payments in Italy were an example of companies supporting an ideologically sympathetic range of parties in a country where the danger of a radically anti-business regime coming to power has been a very real worry to the oil companies as well as to other foreign investors. The companies involved argue that they were merely responding to heavy pressure from Italian politicians. Explicit or implicit threats, whether from a Nixon campaign manager wanting contributions to the Campaign to Re-elect the President, or from President Park's party in South Korea demanding contributions to its campaign chest, can be difficult to ignore. The kind of dilemma presented was expressed by Gulf's chairman, Bob Dorsey, before the Church Sub-Committee when explaining why his company gave $1 million in payments in Korea in 1966:

> The demand was made by high party officials and was accompanied by pressure which left little to the imagination as to what would occur if the company would choose to turn its back on the request. At that time the company had already made a huge investment in Korea. We were expanding and were faced with a myriad of problems which often confront American corporations in foreign countries. I carefully weighed the demand for a contribution in that light, and my decision to make the contribution of $1 million was based upon what I sincerely considered to be the best interest of the company and its shareholders. (Church Report, 1976, pt 12, pp. 9–10)

It is only possible to guess the extent of political payments in the past for the obvious reason that large-scale bribery has been a rare occurrence and it would seem that the bulk of the cases of unusual payments in which the oil industry was involved were in pre-1939 Latin America. There were fewer cases after the war and increasingly, when they did occur, these were the result of extortion by local political figures or of pressure from the US intelligence community.

Pinelo's work on IPC, Jersey's subsidiary in Peru, would support this opinion. He suggests that the 1930s was the decade when the company had most influence in Peruvian politics, but it would appear that this stemmed less from the use of political funding than from the fact that this company was far and away the soundest and largest commercial institution in a country which entered the Great Depression of the 1930s in a virtual state of bankruptcy. IPC became the power behind governments because of its ability to grant loans or advance future export taxes. Requests for payments were carefully weighed in terms of their consequences. For instance, one demand for increased advances was turned down by a local IPC manager 'in view of the precarious nature of the Government, the fact that it might be forced out of office, and that its successor may not take it kindly that a foreign corporation assisted to prolong its existence by conniving with it to deceive the public' (Pinelo, 1973, p. 37).

CORPORATE ISOLATION

The picture of oil companies manipulating host governments by fair means or foul is certainly misleading particularly because of the relative isolation of the companies both politically and geographically. In most cases, the companies were extremely unpopular, with local allies far outnumbered by local opponents, and what saved them for so long was their control of a key industry.

Geographical isolation is so obvious that it is often overlooked. Oil companies, unlike manufacturing industries which naturally locate in large cities, have to base their operations where oil is actually found and at the point on coasts where trans-shipment is most logical. In Venezuela, for instance, the oil industry focused around Lake Maracaibo, well away from Caracas. This isolation from mainstream Venezuela was enhanced when the industry built refineries on the Dutch islands of Aruba and Curaçao, just north of Venezuela in the Caribbean. Abadan was an uninhabited island in the south of Iran, and it was not until the 1920s that Teheran was able to establish unquestioned control over the area in which the oil industry operated. This picture of oil operations isolated from the political centres of host countries holds true elsewhere, as in Saudi Arabia, Iraq, Indonesia, Algeria and Peru. Nigeria is rather more complicated, with oil discoveries made on, or off, the more populous coastal areas, but as northern Nigerians took the political lead in the years leading up to the 1967 Civil War, there was an increasing divorce between the economic and political centres of the country. It has really only been in the tiny sheikhdoms and emirates around the Gulf that the location of industry operations put oil companies in everyday contact with local political elites. This geographical isolation fed the natural inclination of companies to operate within self-contained enclaves which normally had clear demarcation lines to discourage fraternisation between expatriate managers and local citizenry. Social life centred on social clubs which were either explicitly or implicitly segregated and, though it would be untrue to say that managers charged with negotiations with local governments never mixed socially with indigenous politicians, such interaction was rare, at least in the Middle East up to the late 1930s.[6]

This isolation could have been counteracted by strong company representation in host country capitals but, in fact, these offices tended to be small and were given little responsibility. For instance, the Baghdad office of IPC during the 1930s consisted of about ten people who were there primarily to look after mundane matters such as travel arrangements. If anything serious came up, senior personnel flew in from outside the country to deal with the problem. The impression is that companies only started taking representation in host countries seriously in the 1940s and 1950s, when host government pressures started increasing. This change was noticeable in Venezuela in 1942–3 when Jersey restructured its operations by taking firmer control of its local subsidiary Creole, ousted, on State Department advice, its Caracas representative, as being too closely identified with the former Gomez regime, and brought in a new representative to enter into negotiations with Venezuelan officials (Taylor and Lindeman, 1955, p. 89; Wilkins, 1974, p. 271).

Stronger company representations could only marginally hold back the tide of opprobrium which faced the oil companies when public opinion became articulate in host countries. Whatever advantages friendship with a company might offer a regime in terms of a special relationship with a government and a smooth-running oil sector, there seem to have been few individuals or political movements which did not harm themselves, in terms of popular support, by following a tolerant policy toward the majors. For every coup which has benefited the industry in a host country, there has been a combination of riots, assassinations and political defeats meted out to industry defenders which far outnumbered any popular manifestations in support of the companies. For instance, there were the sustained nationalist attacks on the companies in Romania at the beginning of this century and mobs rampaging against oil company installations after the death of Gomez in Venezuela (Lieuwen, 1965, p. 51). There was an era of mob violence aimed against Anglo-Iranian in Iran after 1945 which culminated in the assassination of a prime minister, General Razmara, who argued against nationalisation on pragmatic grounds, and there were riots aimed at companies such as Shell in Indonesia in 1957. Even in Saudi Arabia, where the ruling elite has been on warm terms with the industry, Aramco complained in 1950 that Saudi dissatisfaction was putting American lives in danger (Church, 1974, pt 7, pp. 127, 136). Many would say that two Argentinian administrations, those of Peron in 1955 and of Frondizi, fell largely because they were too lenient to the companies (Odell, 1968, pp. 284–6). There are no examples of mobs running through the streets of a host capital city in defence of the oil industry. Whatever the motivation of the mobs which led to the overthrow of Mussadiq, they were not calling for the restoration of Anglo-Iranian. Although circumstances have dictated that host governments have generally had to accept the companies, when local politicians have 'supped' with the oil industry, they have invariably used their longest spoons. This need to look both ways, and establish a working relationship with the industry without appearing to be industry stooges has led to striking instances of tergiversation, as when one Peruvian administration in the 1930s launched a public campaign against the legality of IPC's status while, at the same time, sending the Finance Minister to ask for secret, badly needed, financial help (Pinelo, 1973, p. 36).

It is obvious why the companies have been unpopular. They have been too large, too obviously foreign, too closely tied to imperial powers, too heavily involved in an emotive extractive industry and too secretive about their operations. Balancing this hostility has been support from a few quarters. The least useful ally has been the local workforce which could sometimes be counted on for support. Wages and fringe benefits from working in an oil enclave in societies with minimal employment prospects and primitive welfare services could obviously prove highly rewarding. But there was inevitably resentment against the filling of best-paid and most responsible jobs by expatriates. Moreover union organisations were often politically hostile to the establishment in control of the host country. At times, a workforce would come out as an ally against economic nationalists. This happened, for instance, in Peru where IPC's workforce consistently opposed nationalisation attempts from 1959 to 1968. But in this case the isolation of oil operations worked against the company, since the workers' political impact on decisions at the centre was minimal (Pinelo, 1973, pp. 56–7).

Another ally has been the indigenous business elite, which has been of more value than the workforce, but has proved just as difficult to rely on. The obvious common interest was resistance to left-wing governments of any sort, since measures like nationalisation and high taxation are double edged and can be used against domestic, as well as foreign entrepreneurs. On the other hand many local businessmen have felt threatened by foreign investment and have therefore supported relatively nationalistic policies. In both Peru and Venezuela, for instance, there was a division of interests between the foreign and indigenous capitalists. In Peru the family controlling the very influential paper *El Comercio* launched a campaign against IPC in the late 1950s, apparently to keep in with leftist, anti-American opinion. After the announcement of the Alliance for Progress, the paper gained the support of other Peruvian vested interests which felt that the USA was promoting reform in Latin America at their expense (Pinelo, 1973, p. 113). A similar phenomenon was found in Venezuela, in 1966, where the companies overstepped the mark by fighting a successful campaign against the oil provisions in a tax reform packet, then leaving the domestic sector to fight by itself once the companies had won their own particular victory (Tugwell, 1975, pp. 88–95, 161–2). All these cases, however, are from Latin America. In the Middle East and many other oil-producing areas, the industrial and commercial sectors were too small to have much influence in their own societies.

The most important company allies have been the ruling elites of the host countries. Despite nationalist pressures from beneath, and probably despite personal prejudices, the vast majority of local elites continued to accept that the activities of the companies should be tolerated at least to the end of the 1960s. In the early days the oil companies were essentially establishing mission posts of advanced industrial practices in primarily pre-industrial societies. The local elite did not have the commitment to industrialisation, the resources or the experience to get the best possible bargains from the companies. Yet it was the success of the latter which was chiefly responsible for catapulting these elites into the twentieth century and, in so doing, the companies laid the foundation for their eventual replacement by local competitors.

STRENGTH OF THE COMPANIES

Finance

In the beginning, the companies had the upper hand principally because they could mobilise financial resources far in excess of host governments' capabilities. The late nineteenth-century negotiations for mineral concessions in Persia, for example, took place against a background in which a trip to Europe by the shah would be a major financial commitment (Elwell-Sutton, 1955, pp. 11, 14). In 1903, the Persian government could still be satisfied by a cash payment of £20,000 and a share in the D'Arcy operations. By the outbreak of the First World War, Anglo-Persian, which grew from D'Arcy's concessions, had been forced to capitalise itself at £4 million, with the British government stepping in to provide half of this sum in the same way as Burmah Oil had earlier rescued D'Arcy when he ran out of money (Elwell-Sutton, 1955, pp. 15–25).

Although circumstances varied from country to country, this kind of financial imbalance was typical, with companies struggling to raise what were then seen as substantial sums, while host governments were content to grant them concessions for sums making up a tiny fraction of the overall investment. The international credit rating of a Middle Eastern ruler was not then good enough to finance such speculative investments, and this probably held true for Latin American countries as well. Even though the latter had a longer involvement with international capital markets, this had been mostly in connection with more developed industries like railways.[7]

However, once the initial investment was made by the companies, the balance of power turned, particularly when oil production peaked as it did in Mexico prior to the 1938 nationalisation or when domestic consumption caught up with indigenous production as it did in Peru in the mid-1950s. Finding money to keep oilfields or refineries running is not so difficult as funding them from scratch. Once the industry was well established, it was easier to boost the share of governmental income per barrel and hence provide an important source of revenue which could increasingly be used to replace investment and plant in some of the sectors of the economy where the majors were active. But finding alternative sources of funding is never easy and was virtually impossible during the Great Depression. After the war, the growing importance of aid meant that efforts by hosts to oust the companies might be met by cuts in financial support from parent governments. It was really only in the 1970s, when successful militancy had boosted host government incomes and the Eurocurrency market had widened, that local authorities could really hope to finance an expanding domestic oil industry from sources beyond the control of the majors.

Exploration and Production Management

Even if the hosts had managed earlier to put together the necessary finance, they would have found it virtually impossible to lay their hands on the requisite technology and manpower, since both of these were effectively controlled by the majors until very recently. The local population generally lacked an adequate general education for higher level management and planning positions, as well as sufficient experience in specialised petroleum technologies. Even in Venezuela, which was one of the more developed host societies, Creole claimed as late as 1952 that it could only find fourteen Venezuelan college graduates who were both available and qualified to take staff jobs (Taylor and Lindeman, 1955, p. 33). It was not until the late 1950s that enough local ex-company employees reached government positions of sufficient importance to embark on a realistic national petroleum programme. Men like the Saudi oil adviser Abdullah Tariki, ex-Texaco and ex-University of Texas, who played an important part in the creation of OPEC, were still sufficiently rare that they automatically gravitated to policy-forming levels. There were not enough men of this kind of experience for host countries to staff a national oil company with local citizens.

It appeared that if the host governments were to break the majors' hold, they would have to work with the Soviet Union and Romania which had a lengthy experience in oil production and were only too happy to embarrass the Western oil industry, or else with Western companies which were independent of the majors. Although the 1920s had seen quite a bit of competition in the oil industry, the slump in the 1930s, the Texas oil discoveries and the wartime devastation in Europe all conspired to leave the majors very much in control. However, there was a re-emergence of activity in the industry's fringe in the late 1940s. This meant that host governments could get parts of the operation of the oil industry performed on more generous terms than those offered by the majors and then, slowly, give their own national oil companies increased experience in the overall management of oil projects, with the aim of developing their detailed grasp of oil operations in time. The terms offered the hosts in 1948 by independents Getty and Aminoil to enter the Neutral Zone between Saudi Arabia and Kuwait were notably more generous than the majors would have preferred. From 1957 Iran was a pioneer in joint ventures with a number of agreements made between foreign companies and NIOC. The initial breakthrough was when the Italian state company, AGIP, and an American independent, Pan American Petroleum Company, each entered into joint ventures and by the 1960s even a major like Shell had become part of such a deal. When Iran went one stage further in the mid-1960s and sought companies willing to operate as service contractors, it was France's ERAP, a consortium of European national champions and the American independent, Conoco, which led the way (Fesharaki, 1976, pp. 70–84). Although the newcomers were not necessarily as effective in their operations as the majors, the mere fact that the number of companies involved in exploration in the Middle East went up from 9 in 1940 to 126 in 1976 increased the bargaining strength of the host governments. It gave them the option of unbundling the package of operations traditionally provided for them by the majors, so that they could carry out some of the activities themselves, only calling in foreign help for the parts that were left (Exxon, 1976, p. 20).

In the end, there was very little that the majors could do to maintain their absolute control of the industry in a country determined to replace them at all costs. On occasion they tried to organise boycotts of labour and services in an attempt to strangle the operations of newly expropriated properties. The Mexicans survived such an attempt in the early 1940s. Although Pemex took more than thirty years to make any major new finds of oil on Mexican territory, it has arguably operated well enough to give developmental gains to the Mexican economy (Tanzer, 1969, pp. 288–303). The companies tried a similar boycott when the Cubans expropriated the refineries of Shell, Jersey and Texaco in 1960, and once again failed, apparently because Russian technicians easily overcame the technical difficulties stemming from shortages of spare parts, with the result that oil products continued to be available for local consumers (Tanzer, 1969, pp. 327–44; Bonsal, 1971, p. 50). Although the Cuban case is a downstream one, concerning refining rather than crude production, Tanzer is undoubtedly right when he claims:

> The fact is that in today's world of spreading technology, no company, group of companies, or single country can realistically hope to maintain a stranglehold on technical skills, particularly one as widespread and long established as oil refinery operation. Because of this diffusion of technological skills, . . . [boycotts] on spare parts [are] also doomed to failure . . . such a boycott takes a considerable time to work since there is usually some stockpile of spare parts on hand and/or improvisations can be made. At its most effective, such a boycott may cause inefficiencies but it cannot usually be decisive. (1969, p. 335)

Transportation and Marketing

Undoubtedly, it has been the companies' ability to deny markets for the oil of countries with which they are in dispute which has been their most effective card, and one whose importance was strengthened as the glut of Middle Eastern oil which appeared on world markets after the end of the Korean War put most of the bargaining power into the hands of the ultimate purchasers of crude oil. The classic case of the use of this card was the embargo against Iranian exports after the 1951 expropriation of Anglo-Iranian. These were cut to a mere trickle to Italy and Japan; the British used their clout in the international legal arena and the other majors rallied round Anglo-Iranian in refusing to deal with Iran and in exerting pressure on independents which might be tempted to step into the vacuum.[8] There had, however, been earlier cases of the use of this device. In Peru, Jersey's IPC cut off supplies of oil to Lima, the capital, in the course of a dispute with that government in 1918 (Pinelo, 1973, pp. 17–19). Likewise, the companies embargoed Mexican oil exports after the 1938 expropriations, but this measure soon became irrelevant when rising Mexican demand led to there being little oil for export anyway.

This device was most useful in the Middle East, particularly during the 1950s and 1960s. Even though Anglo-Iranian/BP no longer dominated the Iranian industry, it was clear that Mussadiq's government had been starved of revenues through the export embargo before the uprising took place against him. The lessons were equally clear. The problem had never been one of producing or refining the oil, for the president of Cities Service reckoned that some $10 million would be enough to get the Abadan refinery back into action. The question was to find available tankers and ultimate markets. Of these, tankers were the lesser problem. Even though much of the world's available tankerage has always been tied to the majors by long-term contracts, the majors' formal hold on this activity has always been the weakest of all the areas in which they are involved. They held only 29 per cent of the tanker fleet in 1953 and only 19 per cent in 1972.[9] On the other hand their control of product marketing and refining activities has been much stronger, for these are activities which, with the exception of plants like the Abadan refinery, have lain outside the grasp of expropriating nationalists. However successful host governments were in finding technicians to help keep the oil flowing, they were lost without customers on world markets. These were difficult to find as long as the bulk of the world's refineries have either been owned by the majors or tied to them by long-term supply contracts, and this difficulty was exacerbated by the general world glut of oil in the 1960s.

Although the control by the companies over market operations gave them their strongest hold over host governments, it is interesting to see how few times this device was actually used. The Iranian case was the last full-blooded embargo of a country's exports, though the apparent slow development of Iraqi exports in the 1960s in comparison with other Middle Eastern producers was a reminder of the weakness of host governments in this area. In practice, embargoes have been a dangerous weapon for the companies, since their use is a declaration of economic war, and the companies have always been vulnerable to retaliation against their property in the host country. The attempts of the majors to reconcile the conflicting demands of Middle Eastern producers for higher production during the 1950s and 1960s show dramatically how vulnerable companies would have been if they did not fulfil production expectations. Host governments certainly realised their marketing weakness in terms of doing without the majors altogether, but some, like Iran, became insistent that the companies should fulfil certain minimum export levels. The companies ignored such pressures at their risk.

Knowledge

None of the above tactics really explains how the majors held on to their concessions for so long. When analysed singly, each tactic apparently open to the majors had been shown to have drawbacks and, in any case, the companies did not really use any of them very much. It was almost as if even potentially militant hosts like Venezuela, Iran or Iraq were frightened of going in for the kill, yet, as events of the 1970s have shown, once the producers banded together they were able to get rid of the traditional system of concessions within a decade and to raise their return from each barrel of crude oil some elevenfold within the five years from 1970. With the possible exception of Iran in the early 1950s, the actions of the majors have not led to the destabilisation of host governments. The companies found few allies within these societies and the embargo weapon was a double-edged weapon. Why didn't the hosts rebel earlier?

Undoubtedly market movements are one explanation. The glut of crude oil on world markets after the mid-1950s meant that the real price of crude oil fell for more than a decade. The lessons of Iraq and Venezuela, both among the most militant hosts, was that oil exports and government revenue would suffer if the companies were crossed. Certainly it was no accident that the final breakthrough came as the market tightened significantly for the first time for over a decade, and in Libya, one country whose exports were particularly significant in the key European market. However, the tentativeness of the initial response by other hosts to Libya's initiative suggests that it is necessary to look deeper into the psychology of host governments.

In fact, whatever the weakness of company defences which is

apparent in retrospect, the host governments did not realise it at the time. Their knowledge of the complexities of the industry was scanty, their experience of serious bargaining with the companies was limited and their awe of the companies was great. As the Shah of Iran, one of OPEC's more militant figures, put it when describing the attitudes of that body's members in the early 1960s: 'I must admit we were just walking in the mist; not in the dark, but it was a little misty. There was still that complex of big powers, and the mystical power and all that magic behind the name of all these big countries' (Sampson, 1975, p. 160). Relative ignorance about the industry's inner workings explains a lot of this feeling of impotence. For instance, Venezuelan administrations from the late 1950s were relatively advanced in trying to get to grips with the industry, but they still had problems in trying to monitor it systematically. As Tugwell puts it:

> Before 1959, Venezuela experts were generally well informed about the technical aspects of oil production within the country, in the sense that they know the principles of conservation and the best procedures for bringing oil to the surface; but they had very little idea of what went on in the industry as a whole. (1975, p. 56)

They therefore tried to influence the companies' pricing policies on international markets through the intervention of a Co-ordinating Commission which actively scrutinised the sales contracts of the companies, but these efforts lasted a mere two years. In Tugwell's words again:

> A major problem was administrative: the Commission was never able to keep close enough track of the companies' marketing decisions. Its leaders, though highly trained, lacked experience, and the agency itself lacked the staff and facilities needed for the complex job it had taken on . . . it found itself on the defensive with fewer facts at hand than the companies . . . Thus, in practice, the government was tagging along after the companies, urging them to do its bidding while fully aware that they knew more about the actual market conditions. (1975, p. 59)

This was not a dilemma unique to the Third World, for industrialised countries like Italy, Germany and Belgium were insisting in the mid-1970s that they too should be given much more information about the workings of the oil companies whose juggling of crude oils from various sources and manipulation of transfer prices between subsidiaries shrouded these operations in sufficient uncertainty that consumer government authorities could not be entirely sure that they were not being charged significantly more for crude imports than some like-placed neighbour. Because of its shortage of experienced policy makers, the average host government was even more badly placed. In some cases, legislation governing oil exploration and production had actually been drawn up on the recommendation of the majors.[10] Even when that was not so, the thought of the complexity of having to launch into international markets without the aid of the majors was enough to daunt all but the most hardy.

What was obviously needed before the host governments could really move forward was some forum which allowed them to draw on each others' experiences with the companies. Even before the formation of OPEC, producer governments did not exist in a total vacuum. In 1918 Peruvian authorities were encouraged by the press to consult the new Mexican laws controlling the oil industry, and the American ambassador in Lima was ordered by his superiors to keep watch for signs of Mexican influence (Pinelo, 1973, p. 20). In 1948, Venezuelans, aware of the need to get Middle Eastern producers as well to demand fifty-fifty profit sharing, translated their documents into Arabic and sent a diplomatic mission to the Middle East to explain the principle. Even though neither Iran nor Saudi Arabia allowed them near their oil officials, they were received in Baghdad and Basra and the message got through (Church, 1974, pt 7, p. 168). When the Libyans drafted their 1955 Petroleum Law, they consulted a group of experts which included Nadim Pachachi, later secretary-general of OPEC. It was these experts who were responsible for insisting that there should be many concessions, so that the majors would have to compete with a number of independents. It was the initial success and the ultimate vulnerability of these independents which was to strengthen the hands of Qadafi at the end of the following decade.[11]

OPEC's importance was that it institutionalised co-operation between host governments which were only sporadically in touch with each other before its creation. In the beginning its impact was limited. Market forces during the 1960s were such that the bargaining position of oil producers, however united, was bound to be weak. All that OPEC members could do was to impress on the oil majors that further cuts in the posted price would be fought, and to start disputing issues like that of royalty expensing, winning the kind of adjustment they wanted and thereby slightly raising their take from each barrel of oil that was lifted. However, for most of the 1960s, OPEC was more a clearing house for ideas than anything else. But that had its importance. The fact that the participation issue was under serious discussion within OPEC in 1969 can only have hardened the resolve of its members in the following dramatic years.

To turn OPEC membership from a loosely knit group of government representatives into the genuinely effective bargaining force of the 1970s needed two things. First, it needed a buoyant demand for oil which would move the industry more into a sellers' market. This happened in the late 1960s when non-communist oil consumption rose at an annual rate of over 8 per cent three years in a row (1968–70) and, in sympathy, the Eastern hemisphere's oil production rose at over 12 per cent per annum during the same years (BP, 1976, pp. 18–21). Secondly, however, it needed a catalyst. The Saudis were not going to lead a confrontation with the industry. Iran was building up its pressures on the companies year by year but had not yet steeled itself to push them to the limit. With the arrival of Qadafi in Libya the scene changed. The rapidly expanding Libyan production was particularly crucial to Western Europe, the structure of the industry within the country meant that there were relatively small companies which were overdependent on Libyan oil and Qadafi was not personally overawed by the reputation of oil companies. Once Libya had broken the ice and shown that the international industry could be forced to concede rises in the posted price of oil, it was as though a log-jam had been broken. The Gulf states, led by Iran, were determined to emulate Libya's example. Eventually, the producers' concern with the price issue was turned to the participation issue, and the very existence of 'foreign' oil companies at the production end was called into question. It was almost as though the mystique of the companies which had served them so well over past decades had gone, thus allowing much more timid and conservative regimes to join an onslaught on the companies which had held them in a state of psychological dependency for so long.

NOTES

1 However, as Keohane points out, these leaders were more concerned with their personal and family needs than with maximising the national interest.
2 The Pearson/Cowdray interests were to sell out to Shell.
3 Some of these allegations came from contemporary diplomats such as US Ambassador Lind who firmly believed that Pearson was behind both Huerta and British policy toward Mexico (Calvert, 1968, pp. 234, 73–84).
4 The invasion was primarily aimed at ousting Reza Shah whose pro-German position was endangering one of the key supply routes to Russia (Longhurst, 1959, pp. 99–101).
5 The Hickenlooper provisions did lead to a cessation of aid to Ceylon.
6 One company executive has pointed out that this was all part of the colonial mentality of the times which affected expatriates in any line of business. Segregation was never the official policy of a company such as BP.
7 One executive points out that the industry has become more capital-intensive as it moves from exploration to production. Traditionally, the latter stage required some ten times the amount of capital needed in the earlier stage. This ratio may now be widening.

8 The American independent, Cities Service, did seem interested in helping the Iranians but its interest evaporated around the time it was given long-term supply contracts for Middle Eastern oil from the majors (Elwell-Sutton, 1955, p. 296; Tanzer, 1969, pp. 417–18).

9 The shipping issue is complicated by the existence of a long-term charter market which the majors may be said to control. Issawi and Yeganeh claim that the majors owned 47 per cent of the world tanker fleet in 1959 (1963, p. 61). Adelman gives the issue more serious consideration and shows that between 1966 and 1968 the majors owned between 17 and 20 per cent of the tankers registered outside the USA (1972, pp. 104–6).

10 For the Venezuelan case, see Lieuwen (1965, p. 48); for Peru see Pinelo (1973, pp. 49–50). Note also that the Saudi tax legislation of 1950 was very much set up with American advice (Blair, 1976, p. 199).

11 Rustow has pointed out that Pachachi presumably had learned from the case of IPC in Iraq.

5
Parent and Host Governments pre-1970

THE WIDER ISSUES

For most of the twentieth century, oil was only one issue linking parent governments with the fortunes of the Middle East. Admittedly once oil was discovered, the value of the area increased, but some of the other issues involved would have been sufficiently important to ensure Western involvement regardless of the existence of oil. The question of Israel, for instance, was to lead to foreign policies which clearly over-rode the best interests of the oil companies in the Arab world.

BRITAIN AND THE MIDDLE EAST

The British had been involved in the Middle East to protect their routes to India from at least the seventeenth century. The traditional enemy was Russia, periodically trying to expand southward through Persia, though the incursions of Napoleonic France also led to diplomatic activity and the Anglo-Persian Treaty of 1800. The building of the Suez Canal strengthened Britain's strategic interest in the area and led to the occupation of Egypt on the initial pretext of the latter's defaulting of the debt and then continued to ensure that British imperial communications were protected from local nationalist attacks (Lenczowski, 1960, pp. 28–9, 612).

The discovery of oil in Persia did not immediately over-ride the traditional range of interests of British policy makers. Oil figured in the Mesopotamian campaign of the early part of the First World War only in a minor way, its importance to Britain being far outweighed by considerations such as the India Office's interest in Baghdad, the need to forestall Russian involvement there and the desire to raise British prestige in the Moslem world after the Dardanelles disaster. Towards the end of the war, oil had made more of an impression on the Imperial War Cabinet, and the occupation of Mosul (where oil was expected to be found) which took place after the armistice was signed was done to strengthen Britain's bargaining position with both the Turks and the French (Kent, 1976, pp. 118–19, 126). In the postwar era the British were concerned not only with keeping an eye on oil interests but with trying to arrange a suitable dismemberment of the Ottoman Empire. It was also necessary to try to reconcile the Balfour declaration on Palestine with the promises made to ex-allies in the Arab world.

It is difficult to determine where oil fitted in. Foreign and popular opinion believed that it played a primary role in Whitehall's thinking, but key policy makers such as Lord Curzon, Foreign Secretary from 1919 to 1924, protested strongly. At the time of the 1922 Lausanne Conference he insisted, with regard to the status of the allegedly oil-rich Mosul province:

> The question of oil in the Mosul vilayet has nothing to do with my argument. I have presented the British case on its own merits quite independent of any natural resources there may be in the country . . . During the time I have been connected with the foreign affairs of my country I have never spoken to or interviewed an oil magnate. I have never spoken to or negotiated with a single concessionaire or would-be concessionaire for the Mosul oil or any other oil. (Shwadran, 1973, pp. 219–20)[1]

British concern with routes to India remained strong throughout the 1920s. As late as 1928 a treaty between Britain and Persia was principally concerned with Persia's insistence that foreigners should not have extra-territorial rights and Britain's request that Imperial Airways be allowed rights of overflight over Persia en route from Cairo to Karachi. Oil issues merited only a cursory mention (Toynbee, 1929, pp. 351–6). It was, however, impossible for the British to ignore oil completely, given the Anglo-French and Anglo-American negotiations concerning the final composition of IPC. There were contemporary political figures who insisted on the importance of oil, though, with the retirement of Lloyd George, none of these was of outstanding influence. Shwadran, who is one of the few authors to see oil interests in a wider context, argues that London's policies were more oil-tinged than not, especially over matters like Mosul, but his arguments are not entirely convincing. Certainly, the oil prospects of Iraq and its neighbour, Persia, were widely discussed at the time, but, given the break-up of Turkey's Arab empire in the aftermath of the First World War, it was certain that the victors would become involved in the creation of new states. The Turkish collapse caused a vacuum between Britain's foothold in Egypt and India. It was inevitable that Britain, as an imperial power, would have moved into this area, whatever the richness of its mineral resources.

During the Second World War another non-oil issue became important – the need to keep open the most convenient supply route to Russia. The Iranian leader, Reza Shah, who had come through the 1931–3 confrontation with Anglo-Persian without too much damage, was pressured into abdicating because of his sympathetic approach to the German cause.

After the war, the British position hardened as oil came to play a much more unequivocal role in policy formation. The 1951–4 affair in Iran came at a time when Britain had started to withdraw from its role as world policeman, and the decision to take a hard line in Iran was primarily a function of the size of the oil stake. But there was also the psychological trauma of the decline of the imperial dream. The torrent of abuse against Mussadiq was far in excess of anything demanded by the oil issue. Hysteria unrelated to oil interests can also be seen in the British response to Nasser in Egypt. Oil was a factor in the British decision to join the French and Israelis in the 1956 invasion of Egypt, but it was by no means the only cause.

Worry about oil supplies is the simplest explanation of why the British took part in the ill-advised invasion of Egypt. The Prime Minister, Anthony Eden, was certainly aware of oil and mentioned in his memoirs his fears that nationalisation of the Suez Canal would cut Western Europe off from Middle Eastern oilfields. He felt that Britain could not allow Nasser 'to have his thumb on our windpipe'. Four months prior to nationalisation Eden had put the issue succinctly, when he warned the Russians that the uninterrupted supply of Middle Eastern oil was 'literally vital to our economy . . . we would fight for it . . . we could not live without oil . . . we had no intention of being strangled to death' (1960, pp. 358, 424–6). He then quoted a leading article from *The Times* of 1 August 1956 which put the issues as follows:

> The first [issue] is quite simple. Freedom of passage through the Suez Canal, in peace or war, is a prime Western interest . . . The second issue is no less obvious. The great oil works and fields of the Middle East are one of the main foundations of Britain's and Western Europe's industry and security. Anyone who thinks that a victory for Nasser would not encourage other extremist demands against the oilfields – and against strategic bases – should confine himself to tiddleywinks or blind man's bluff. (p. 441)

Eden had read and been heavily influenced by the pessimistic OEEC Hartley Report which painted a very gloomy picture of the future of Europe's energy supplies. On the other hand, the British government was concerned with other issues as well. Nasser was seen as the principal opponent of Britain's Middle Eastern policy, behind Jordan's refusal to join the Baghdad Pact, the sacking of General Glubb and the increasing opposition to the pro-British Iraqi regime. Eden, failing to understand the nationalist forces sweeping the Middle East, viewed Nasser as a fascist dictator who, like Hitler, must on no account be appeased (Eden had resigned as Foreign Secretary before the war over the appeasement of Hitler). When the canal was nationalised, there were frequent comparisons to Hitler and the Rhineland crisis.

A personal hatred of Nasser came to dominate Eden's thinking. As early as March 1956, before the canal issue had come to the forefront, Eden said to Anthony Nutting, Minister of State in the Foreign Office: 'What's all this nonsense about isolating Nasser or neutralising him . . . ? I want him destroyed, can't you understand?'[2] Nutting concludes that this was 'the last dying convulsion of British imperialism' (1967, p. 12) as Eden was clearly looking for a pretext to destroy Nasser before the nationalisation. In fact, when the invasion took place, the canal was operating efficiently and the threat to West Europe's oil supplies was receding fast. Hence the danger to the oil-flows was clearly a pretext for a policy less concerned with the oil industry (the USA had the same interests and was willing to live with a nationalised canal) and more from a desperate fear that Britain's political hold on the Middle East was being threatened by Nasser. Many oil company executives were alarmed by the invasion which they felt would have the effect of jeopardising British commercial interests in the Arab world (Sampson, 1975, p. 138).

There has been no other outstanding international episode since 1956 in which the interests of oil have been paramount in official British foreign policy. The Nigerian Civil War (see pp. 75–7) had oil considerations in the background, but apart from some hesitancy in June and July 1967 when the renewed closure of the Suez Canal made the British government more than usually sensitive to threats to alternative supplies, the Foreign Office was adamant that Nigeria's territorial integrity should be preserved lest a precedent be established for tribal tensions to be judged an adequate reason for dismembering newly emerging ex-colonial states. British policy was primarily concerned with ensuring that the Soviet Union and France should not gain from meddling in what had traditionally been a British preserve.[3]

Britain's decision to withdraw its forces from the Gulf in 1971 was an acknowledgement that the calls of empire could no longer be met through the presence of troops in the Middle East. This recognition of Britain's economic weakness meant that the oil companies were now demonstrably on their own. It may be that this withdrawal contributed to OPEC's onslaught against the companies, but it may also have helped the latter in the long run by reducing the risk that commercial disputes would be aggravated by political confrontations between host governments and British authorities. Oil diplomacy could take place in a more neutral atmosphere.[4]

THE USA AND NON-OIL ISSUES

Until the Second World War, American interest in the Middle East was a case of flag following trade and the Open Door battles with the Europeans in the 1920s were about commercial self-interest. With the election of Franklin D. Roosevelt, oil issues in Latin America were subordinated to wider considerations such as the Good Neighbor policy which could also be seen as enlightened commercial self-interest. The war, however, strengthened American interest in the Middle East. Events in 1943–4 culminated in the De Golyer Report which made it abundantly clear that the oil industry's centre of gravity was moving eastward and that the USA was badly placed to take advantage of this shift. Since this awareness came in wartime, the importance of the Middle East impressed the military minds of the Pentagon which had to face the fact that Middle Eastern oil must from henceforth figure in strategic planning. The Pentagon's wider interest in the area was to grow when the Joint Chiefs of Staff decided in 1943 to build an airbase at Dhahran in Saudi Arabia as a link between Cairo and Karachi. This was originally planned as a key link in the logistical chain joining the European and Japanese war theatres. In the Cold War it was to give the Pentagon a continued interest in Saudi Arabia's fate (Lenczowski, 1962, p. 552). The late 1940s however, introduced two new themes in US policy which were to over-ride the commercial interests of the oil industry – the Palestinian problem and the Cold War.

Israel

There is little evidence that the oil companies had any clearly formulated position on the creation of the state of Israel, though oil men in the field knew that it would complicate their relations with the Arabs.[5] Quite clearly, the US government, which had taken over the Zionist mantle from the impoverished British, was paying scant attention to oil interests and the majors did not have the political clout to put up any effective resistance, even if they had wished to do so. In many ways, the situation was reminiscent of the import quotas battle of the 1950s, when the domestic arm of the US oil industry was able to get its own way. In both cases, US policy makers were reacting to the pressure groups which could deliver voters to the polls. However powerful the majors might appear in the international arena, they counted for little in the domestic scene. Votes from oil rigs out in the Arabian desert have not swung the fate of American elections.

Oil men in Washington

The decade or so after the ending of the Second World War was one in which businessmen were particularly close to Washington, reflecting their extensive co-operation in the war effort. Naturally there was some influx of company personnel into the foreign policy establishment, for such reasons as Secretary of State Forrestal explained:

> What I have been trying to preach down here is that in this whole world picture, the government alone can't do the job; it's got to work through business . . . that means that we'll need to, for specific jobs, be able to tap certain people . . . (Kolko, 1969, p. 22)

Oil men moved into the foreign policy making departments and diplomats moved in the reverse direction. There was Robert B. Anderson, a Texan oil producer who moved to the Navy, Defense and Treasury Departments, Herbert Hoover, Jr, a director of Union Oil who moved to the State Department, and, of lesser importance, Andrew Ensor, who moved from Socal to the State Department and then on to Mobil. There was also George McGhee, who moved from an oil background to the State Department and then to Mobil's Board. John J. McCloy, who was in the Department of War, President of the World Bank and High Commissioner in Germany, became a leading advocate of oil industry concerns during the 1960s and 1970s. And Walter Levy switched from a position in the Marshall Plan's European Co-operation Administration and in Hoover's team in Iran to a role as an influential petroleum consultant (Engler, 1961, pp. 310–12).[6] Hoover was probably the most important as far as the industry was concerned as he mediated between the companies and Venezuela after General Medina demanded a revision of oil contracts, and played a key role in putting together the industry alliance which permitted a solution to the Iranian crisis of the early 1950s (Sampson, 1975, pp. 109, 128–30). Anderson was less obviously involved with oil, reaching his peak government post as Secretary of the Treasury, but even in his official withdrawal from Washington he was involved in US-Middle Eastern politics, arranging meetings between US industrialists and Nasser in the vain hope that the breach between Cairo and Washington could be healed (Copeland, 1969, p. 233).[7] Levy has been most narrowly concerned with oil politics, especially the Iranian situation.

To prove that the oil industry was a particularly influential group in US foreign policy making, the case will have to be demonstrated in the career of McCloy and perhaps also those of the Rockefellers. In Galbraith's words, McCloy was the 'undisputed chairman of the

American establishment' (1976) for much of the 1950s and 1960s, holding posts such as the chairmanship of Chase Manhattan, the Council on Foreign Relations and the Ford Foundation, and he has obviously been heavily involved with oil industry affairs as evidenced by his presence at the initial meeting of the Petroleum Reserve Corporation during the Second World War, his lobbying of the Department of Justice from the 1960s on the need for antitrust clearance for oil company co-operation, and his central role in the formulation of a common industry position in the 1970s. But his interests have been far from narrowly sectional. He was also Assistant Secretary of War, High Commissioner for Germany, Chairman of the USA's General Advisory Committee on Disarmament and a member of the presidential commission investigating the Kennedy assassination.[8] He was close to the corporate sector in an era when the American business and financial elite was involved in policy making. Whether he spent a disproportionate amount of his time pushing the specific interests of the oil industry and just how successful he was in his attempts is unclear. For instance, despite McCloy's awareness of US vulnerability to the host government threat, the Nixon administration was very slow to see the problems caused by growing US import dependence. There are many reports of frustrated industry delegations, such as the one which included Robert Anderson, John McCloy and David Rockefeller that called on Nixon shortly after his 1968 election victory to urge a more even-handed approach to the Middle East. As one of delegation said: 'We could always get a hearing, but we felt we might just as well be talking to that wall' (Sampson, 1975, p. 206).[9]

There is no doubt that there was an informal foreign policy group surrounding the Rockefellers, the Chase Manhattan Bank, the leading foundations and the Council on Foreign Relations. David Rockefeller, for example, developed an interest in the Middle East and became a useful link between Washington and various Middle Eastern capitals. Nelson Rockefeller, on the other hand, only showed an active interest in Jersey Standard's oil activities at the start of his career; since then, his career almost suggests that 'he's been washing the oil off'. But although this group was undeniably aware of the growing pressures on the industry, there is no evidence that they were able to 'tilt' US policy in any way which significantly helped the companies maintain their privileged position in the oil-producing world.

The domestic arm of the oil industry has obviously had considerable influence. President Lyndon Johnson was strongly beholden to the southern oil industry and had long been a key Congressional figure defending this group from predatory tax reformers. But the international arm of the industry lacked such influence. The relationship between the State Department and international oil companies does not begin to compare with the cosy links between the Pentagon and the various armaments suppliers. In any case, it has rarely been industrialists who have mattered in Washington, but rather lawyers and financiers. It was from these latter ranks that the men of real power came – people like Dean Acheson, John Foster Dulles, James Forrestal, Douglas Dillon, Averell Harriman and Robert Lovett. Even within the ranks of the industrialists who made it to the top in Washington, there have been no oil men with the stature of the two Charlie Wilsons (from General Motors and General Electric) or Robert McNamara from Ford. The picture is one of an industry whose influence on foreign policy has been rather less than its size would indicate.[10]

The Cold War

After the question of Israel, the Cold War was the second major issue which cut across the oil one when US policy toward the Middle East was discussed. Despite the arguments of revisionist historians that the Cold War was caused less by Soviet policies than by American determination to ensure that the Soviet Union did not close off large parts of the post-1945 world to US commerce, there is little evidence that the oil industry entered into the debate. There was only one oil-related crisis which contributed to the build-up of the Cold War, even though the early days of the Cold War coincided with the expansion of US oil interests into the Middle East and thus into an area which was right up against the southern belly of the Soviet Union. In fact, it was in the 1945–6 dispute in this area, when the Soviets refused to withdraw from northern Iran, which was to be the first significant postwar skirmish in the growing hostilities between East and West, which, after Russia's demand for joint control of the Dardanelles and Britain's admission that she no longer had the strength to prop up the regimes in Greece and Turkey, were to lead in 1947 to the Truman doctrine and Marshall's Harvard speech, which both acknowledged that the world was now divided into two competing blocs.

Although Truman's memoirs refer to the 1945 Russian threat to Iran as endangering the Western economies and the raw material balance of the world, there is surprisingly little mention of specific economic factors in the period leading up to the 1947 break between East and West. Uranium came in for some discussion, but the issues of oil and the Middle East were rarely raised. Iran was mentioned but nearly always in the same breath as Greece and Turkey, thus suggesting that Soviet designs on Iran were just another case of Soviet expansion rather than a direct threat to a specifically vulnerable industry. In fact, the nearest oil and the Middle East came to being given a central role in this debate was in an early draft of the Truman doctrine from which Acheson deleted a specific reference to

US interest in Middle Eastern natural resources (Gardner, 1970, p. 220).

Studies of the business community's impact on the development of Cold War thinking are notable for their omission of oil company executives. This contrasts with the mention of Henry Luce, whose publishing empire had been used to swing US public opinion into the Second World War and then to warn America of the dangers of Soviet ambitions in Eastern Europe (Gardner, 1970, pp. 22, 53, 160), and of the industrialist, Alfred P. Sloan, chairman of the board of General Motors, who was particularly active in 1945 trying to persuade Bernard Baruch of the necessity for regenerating rather than dismantling the German economy. It was a top executive of General Motors who asked permission of Forrestal to distribute to supervisory personnel copies of George Kennan's paper which later became the famous 'X' article in the July 1947 issue of *Foreign Affairs*, 'The sources of Soviet conduct'. Others who made contributions to American policies towards Germany, cockpit of the Cold War, were executives from US Steel and Johns-Manville. The oil companies were once again absent (Gardner, 1970, pp. 250–1, 261–2, 283).

It was not until 1950 that the Middle East's implications in the Cold War really hit the State Department with some urgency. Saudi and Kuwaiti production were then in full stream, literally fueling European and Japanese recovery, while US demands for protection against cheap Eastern hemisphere oil were starting to be heard. The Korean War's outbreak coincided with an upsurge of militancy on the part of Middle Eastern host governments. A key policy paper on Middle Eastern oil produced by the State Department in September 1950 starts uncompromisingly:

> 1 Problem
> The Threat of Communist aggression is increasing. The Middle East is highly attractive and highly vulnerable to this threat. Rupture of the flow of Middle Eastern oil to normal markets or seizure of those resources from without or within would seriously affect US and allied economic, political and strategic interests. (Church Report, 1974, pt 7, p. 122)

It was decided that the flow of funds through oil companies to Middle Eastern governments must be increased if anti-Western unrest was to be curbed. One possible solution raised in the State Department paper was that of granting foreign tax credits by the US Treasury so that the burden for buying peace in the Middle East would be shouldered by the US taxpayer. This was the solution which the National Security Council eventually put to the tax authorities. It was a clear case of a decision on oil matters being taken substantially for Cold War motives.

Nasser and the US Majors

In general, the oil industry and the State Department have agreed on policies concerning the Middle East. Where tension has arisen, it has normally come where companies have been unhappy with the amount of diplomatic backing given them in cases of expropriations, or where foreign service officials have felt it necessary to distance themselves from particular company excesses.[11] Probably the nearest to a genuine difference of opinion on wider foreign policy was the issue of how Nasser should be handled.

Miles Copeland argued that the 1958 Lebanese crisis hardened corporate feelings against Nasser somewhat in advance of the change in general attitudes within the State Department. The companies blamed Nasser for triggering a civil war which, by threatening Beirut, endangered the lives of Western businessmen to an extent which no amount of radicalism confined to Egypt could do. Although the US government resisted any conspicuously anti-Nasserite policy for the next four years, Copeland claimed: 'At the same time, the great Washington lobbies of the business world have hammered hard on anti-Nasser themes, and have harassed and denounced "pro-Nasser" officers in the State Department (including even those who insist mildly that Nasser "is not *all* bad") into silence or into asking for transfers' (1969, p. 207). This was despite the fact that the oil company executives most closely in contact with Nasser were privately proclaiming that: 'they would prefer to deal with top level Egyptians even when they were in a hostile frame of mind than with any other Arab leaders when they were friendly' (Copeland, 1969, p. 220). This argument is supported by John Badeau, longtime president of the American University in Cairo who was ambassador to Egypt during the early days of the civil war in the Yemen, a struggle which polarised the US foreign policy establishment. It was a war between a medieval, outmoded imamate, backed by the monarchs of Jordan and neighbouring Saudi Arabia, and a republican movement backed by Nasser. In itself, Yemen was of no particular strategic importance, and the main goal of diplomats was to keep the conflict localised. Once Nasser became involved, however, the situation changed because the Saudi regime felt bound to intervene as it feared that a republican victory in Yemen would destabilise the Saudi monarchy. The policy of the Kennedy administration was to recognise the republican regime while trying to act as mediator – a policy which worried those in the State Department who were concerned about the Saudi monarchy. As befits an ex-ambassador, Badeau's account is suitably circumspect, but it is clear that he views the oil companies (hostile to an appeasement policy which might threaten Saudi stability) and the Israeli lobby (against any conciliatory approach to Nasser) as the two main opponents of US policy. Basically, the tension was over tactics as much as anything else, since Washington's recognition of the republican regime stemmed from one over-riding concern which both sides held – that the stability of the Saudi monarchy came before all else. Washington believed that an early recognition of the Nasser-backed regime might keep a potentially explosive dispute localised. The companies appear to have been more hard-nosed on the issue – any concessions at all to Nasser were wrong since they would inevitably endanger stability in Saudi Arabia (Badeau, 1968, pp. 123–51).

All in all the picture which emerges is one in which neither oil nor the oil companies have made a very significant impact on US policy makers. In official US government ideology, diplomatic aid to companies establishing themselves abroad and protection to them once they were established was perceived as a legitimate, but not over-riding, diplomatic task. This meant that the foreign policy establishment became quite heavily involved with oil problems on occasions, depending on the physical distribution of sources and the political goals of the authorities claiming jurisdiction over these deposits. In general, access to oil has not been for the first two-thirds of this century a particularly important US foreign policy goal. Much of the interaction which has taken place between the oil companies and the Washington establishment since the Second World War has taken place not because the companies have merited any special treatment, but because they have been key economic actors in the Middle East. Obviously the relative importance of this area in world affairs has been increased by the existence of oil. It is difficult to imagine the State Department becoming embroiled in the Yemeni Civil War had it not been for the dangers posed for the stability of the oil-controlling Saudi regime. But this does not mean that the companies dictated this involvement. Since Saudi Arabia was an important supplier of oil, the USA would have intervened regardless of the form of corporate involvement (the USA had entered the earlier Iranian dispute with Anglo-Iranian even though it was a British company which was in jeopardy).

There was generally a complementarity of interests between US corporate and government officials. Both were interested in seeing plentiful and secure supplies of oil flowing within the non-communist world and both preferred to deal with stable, pro-Western governments. There have been few cases where the interests of the majors and the State Department have been radically distinct. One classic instance was the creation of the State of Israel. Another was in 1973 when the Aramco partners became aware that Saudi Arabia was about to react to the lack of American progress in resolving the Palestinian problem and were unable to influence State Department policy.

The majors have been effective in looking after their own interests in subjects such as taxation, but for the foreign policy machine they have just been one of many concerns. The State Department has found them a useful set of instruments to be manipulated in the Middle Eastern juggling act, but there is no evidence that the oil companies have managed to 'tilt' Washington's policy as far as the area as a whole is concerned.[12] Perhaps, as Rustow suggests, it might have been different if the USA had already in the 1950s been some 40–50 per cent dependent on Middle East imports. As it was, the potential impact of oil on Europe and Japan had most effect on US policy makers.

THE DUTCH AND THE FRENCH

This picture of oil companies being kept well subordinated by foreign policy establishments seems, in general, to hold true for the two non-Anglo-American parent countries, France and the Netherlands, though the detailed historical evidence is considerably less voluminous.

The Dutch have been relatively inactive in non-European oil politics although Indonesia's search for economic as well as political independence meant that, until Shell decided to pull out in 1965, Dutch relations with its ex-colony were inevitably complicated by the existence of oil, as well as by other politically sensitive investments. Dutch oil policy had turned inward on Europe some time before this, a decision which logically stemmed from the fact that Rotterdam had become the leading refining centre of postwar northern Europe. This concentration on the immediate neighbourhood was strengthened by the 1959 discovery of the Groningen gas deposits, whose natural export markets were immediately over Dutch borders. Beyond Europe, the Dutch have followed policies which seem to ignore the implications for their oil interests. The institution of an arms embargo during the latter part of the Nigerian Civil War, motivated by humanitarian concerns, antagonised the ascendant Federal authorities and could well have endangered Shell's extensive interests in Nigeria. The 1973 oil embargo against the Netherlands is clear proof that Dutch policies on the Israeli issue were hardly formulated with prime consideration to economic self-interest.

French policies have been more complex. France's dependence on imported oil has provided a motive for fighting to maintain, if not expand, a marginal role in the international oil industry in the Middle East. There have been times, such as from the last part of the First World War through to the mid-1920s, and the period when the Red Line Agreement was dissolved in the late 1940s, when French diplomacy was actively concerned with oil matters. However, it is important to put the French oil motive in a wider context. France felt it had prior claims to the areas roughly covered now by Syria and the Lebanon and, during the 1920s, French rivalry with the British stemmed at least in part from the very different concepts which these two powers had formed about how the Turkish Empire should be administered (Shwadran, 1973, p. 215). The French were obviously in no position to take part in the Anglo-American oil diplomacy during the Second World War, and, in the war's aftermath, could only fight a rearguard action against the scrapping of the Red Line Agreement. By the 1950s, French ambitions in the Middle East were severely constrained by the illwill within the Arab world generated by France's resistance to Algerian demands for independence. In addition, French involvement in the ill-fated 1956 Suez War, motivated partly by resentment over the role that Nasser was playing in Algeria, meant that France's relations with the Middle East were strained for some years to come. The Algerian dispute itself was affected by oil as the French badly wanted to salvage their hold on the Saharan deposits regardless of the political settlement, as Algeria was the most promising source of crude oil outside the control of the Anglo-Saxons.

The position of oil within Gaullist foreign policy is particularly opaque. Not only did the General launch a campaign which was relatively pro-Arab, he involved France in negotiations in Iraq in 1967 which were seen as particularly unhelpful by the majors. French intervention on behalf of the Biafran secessionists in the Nigerian Civil War was commonly seen as being motivated by oil. And in the background were France's economic relations with Algeria. But although France sometimes stood to gain improved access to oil by its rejection of Anglo-American leadership, the importance of oil in French foreign policy should not be overstressed. De Gaulle's challenge to the Anglo-Saxons rested on motivations which were much more complex than mere economic calculations. The recognition of China in 1964, the break with the USA over Vietnam and the decision to launch a distinctive campaign aimed at increasing French influence within the Third World in the winter of 1964–5 were all motivated by a desire to re-establish France as force in world politics. The improvement of relations with the Arab world should be seen primarily in this light, with any gains in access to oil a pleasant, but not essential, side-benefit. Similarly, the intervention in Nigeria should probably be seen as an attempt to extend French influence in the ex-British African colonies, rather than as a simple grab for Biafra's oil. Oil has been a factor in French foreign policy, but rarely a dominant one.

INTERGOVERNMENTAL RELATIONS AND THE COMPANIES

It would seem logical that the foreign policy of a host government would be principally concerned with the country which was the parent of oil companies having concessions in its territories. For instance Saudi diplomacy should be expected to be principally concerned with the USA from the moment that Socal, and not IPC, was granted the original concession. However it is important to remember that it was generally not a historical accident that companies of a particular nationality came to be active in any one country. The fact that Anglo-Persian came to dominate Iran and to have a leading role in Iraqi and Kuwaiti developments was a direct result of Britain's hegemony over the area at that particular time. Moreover, it was not accidental that King Abdul-Aziz was interested in doing business with an American company, as he wanted to avoid depending on the British who had not proved themselves good friends of the Saudi dynasty.[13] It was hardly accidental that Latin American oil development was primarily in the hands of American companies (though Shell's role should not be forgotten) since this continent was very much within the American sphere of influence. And even though a company like Shell was able to break into Venezuela and Mexico, despite the prevailing patterns of Great Power interests, its Dutch and British connections made little impact on the development of the diplomatic patterns of the host countries. There is little evidence that Venezuela became sensitive to the wishes of Holland and Britain, and, with the exception of the breaking of Anglo-Mexican diplomatic relations over the 1938 expropriations, neither did Mexico become sensitive to European concerns.

The existence of oil has brought particular countries to the attention of the Great Powers. The patterns of the diplomatic relationships of such countries came to be determined less by the nationality of the companies granted the original concessions than by the nationality of the Great Power dominant in that area of the world at that particular moment. The presence of Shell has been largely irrelevant to the diplomacy of Venezuela and Mexico because both these countries are within the US sphere of influence and no powers have challenged US hegemony in Latin and South America. On the other hand, the pattern of recent Iranian diplomacy has been relatively independent of Britain, despite the fact that Anglo-Persian dominated Persian oil production for many decades. The nationality of the relevant concessionaires changed and the focus of Iran's diplomacy reflected the global influence of the United States.

In looking at the industry, the nationality of the majors is of little use in predictions. The key factor is the global balance of power. The nationality of the dominant oil companies in a given area at a given point in time merely gives an insight into the international pecking order some ten years earlier. It has never permitted predictions about which companies will gain at the expense of others in the future.

NOTES

1 Shwadran believes that Curzon and the British protested too much, but a case can be made in their defence. The Turks apparently would have been happy to grant the British mineral rights providing that Mosul was ceded to Turkey. The British turned this compromise down. Shwadran argues that there was so much world attention on this issue that the British could not have defended such an obviously mercenary deal (pp. 227–31).

2 His analysis that the implicit decision to destroy Nasser had been taken well before the canal nationalisation is supported by Childers (1962, p. 187).

3 Keohane argues that the emphasis the British have placed on relations with Nigeria must stem partly from the importance of Nigeria's oil.

4 Rustow and Mugno argue that the Gulf states would have been somewhat less enthusiastic about following Qadafi's example if the British Navy and Air Force had still been active in the area (1976, p. 17). It has also been pointed out to me that I have probably played down the role of oil in Britain's concern with the Buraimi Oasis, the dispute between Iraq and Kuwait in 1961 and the problem of insurgence in Oman.

5 Lenczowski mentions *Nation*, 1948 (1962, pp. 396–7). For the views of an oil man out in Saudi Arabia at the time, see Cheney (1958, pp. 80–6).

6 Other names mentioned include Emilio Collado, who moved from a brief stint in the State Department to Exxon; Melvin Conant, who moved from Exxon to the Federal Energy Agency; William Eddy, who became a consultant to Aramco after having been Resident Minister to Saudi Arabia; Harold Minor, who was Ambassador to the Lebanon and moved then to Aramco. These cases indicate that there is middle-ranking movement between US officialdom and the oil industry, but none of the individuals concerned were important policy makers.

7 Anderson did play an important part in apparently stiffening the resolve of the majors against Castro's demands in 1960, and he was closely involved with the setting up of the mandatory import quota scheme of 1959 (Blair, 1976, p. 173).

8 See Church Report for a full *curriculum vitae* (1974, pt 5, pp. 59–60).

9 On the other hand, Rustow points out that Nixon did send Scranton to the Middle East immediately before his inauguration to fly the 'even-handedness' kite on the Allenby bridge. The industry approach may thus have had some effect.

10 This is not to deny that the oil lobbies have been very effective in seeing that regulatory and tax provisions were drawn up in the industry's favour (see Blair, 1976, for a fuller development of this argument). Nor would I totally rule out the argument that certain law firms are little more than extensions of the international oil industry; however, until this is actually demonstrated, I would prefer to regard them as general representatives of corporate America. It is instructive to note that Domhoff – one of the leading critics of the corporate dominance of US

foreign policy making – barely mentions the oil industry *per se* (Domhoff, 1969).

11 The State Department, like other foreign policy establishments, has had strong motives for keeping on good terms with the oil companies since the latter's men in the field are generally far better placed to get the mood of a country than diplomats who tend to be restricted to the national capital.

12 One non-strategic factor the US policy makers presumably considered is the contribution the companies have made to American balance of payments.

13 It should be stressed that this did not stop him giving the British-dominated IPC a concession in the late 1930s. He was certainly not averse to assuring Britons of his distrust of Americans.

6
The Companies as Transnational Actors

The majors have concerned themselves primarily with the business of extracting and marketing oil and their influence in the political world was originally due largely to the inexperience and timidity of host governments with which they dealt. As the latter increased in self-confidence, the majors could only fight a rearguard action. If these companies survive into the future, it will be because they can perform important economic functions more efficiently than governments, state companies or commercial competitors. If they cannot do this, they will become extinct like all unfit organisms. And it is difficult to make tidy generalisations about how their role has evolved. The claim that the *Pax Britannica* was the golden age for transnational actors overlooks the fact that governments have been involved with the oil industry from the moment when its military importance was fully appreciated in the early days of this century. The conflicting claim, that transnational relations have been growing in importance since the First World War, does not take into account the fact that the increasing economic importance of oil has forced even the most *laissez-faire* consumer governments to become increasingly involved with the workings of the companies.

IMPACT ON THE POLITICAL SCENE

It is obvious that the growth of the oil industry and of the oil companies has had an impact on the distribution of political power in the world. For one thing, the discovery and exploitation by oil companies of sources of crude oil inevitably increased the relative importance of oil-producing countries in the world system. This effect was enhanced by the majors' success in developing world markets. Even when foreign policy establishments might try to downplay the importance of a given oil-producing state, the presence of the majors was a stark reminder that Venezuela, Indonesia, Nigeria or Saudi Arabia had an economic importance which raised them above other nations of similar size, but possessing no oil. Secondly, the oil industry has helped to create nations which, when compared with other Third World countries, have relatively well-developed state bureaucracies. Edith Penrose emphasised the importance of the state sector in a short paper which mentions the fact that as early as 1953 nearly 10 per cent of the Kuwaiti population were in the civil service. Tugwell's study of oil in Venezuela – a society more complex than most Middle Eastern ones – shows how democratic parties and a state apparatus developed, owing relatively little allegiance to the business world, be it in Venezuelan or foreign hands. In fact, their attitude is not just one of indifference to the private corporate sector, but one of positive distrust (Tugwell, 1975). There have been paradoxes. Often the only significant source of tax revenue within the host economy, the oil companies have played an important part in the creation of their own nemesis – a strong state sector with sufficient self-confidence and control to limit or eliminate the role of foreign enterprises. Yet if the state replaces them with a national oil company, it is creating a body with the resources and legitimacy to escape, in its turn, the strict control of the state apparatus. Pertamina, for instance, became virtually a state within a state until overambition led to problems in its international financing in 1975, thus necessitating a rescue by the central government of Indonesia.

The implication of the creation of centralised, well-financed host governments is that the integration of these hosts into the world system may be accompanied by a further attenuation of the role of transnational actors such as the majors. These relatively monolithic governments can be expected to give a very restricted amount of strategic independence to most private organisations operating within their boundaries and private companies, such as the majors, may have a growing problem to win acceptance in societies dominated by functionaries who believe strongly in the economic role of the state. If they do survive in such societies, it will be as a result of their historic ties and hold on technological expertise and not because they are willingly accepted within the prevailing ethos.

SYMBOLS OF THE WEST

Impact on the host societies goes deeper than strengthening the role of central governments. The majors are symbols both of imperialist power and of advanced, capital-intensive commercial enterprises. Their imperial connections tend to repel the elite of host countries who are, nevertheless, fascinated by the life-style which the capitalistic system seems to promise.

The continuation of the concessions system in a period in which nations of the Third World have been casting off the ties of political dependence imposed upon them in previous centuries has created problems. The continued presence of the majors appeared as a reminder that political independence did not guarantee economic independence and the forces which had struggled for political independence focused their attack on what they perceived as economic symbols of imperialism. It is clear that at least part of the effectiveness of the anti-Western campaigns of Nasser and the Soviet Union in the Middle East was their insistence on the oil industry being representative of the Western forces which were allegedly the cause of all the area's misfortunes. An educated Saudi working for Aramco in the 1950s, already aggrieved about lack of promotion prospects, was bound to be receptive to propaganda such as :

> The oil monopolies of the US and Britain have in their struggle for oilfields and pipelines turned the countries of the Middle East into an arena of intrigues, provocations, political murders and coups d'état. The bloody events in Iran, Yemen, Syria, the Lebanon and Trans-Jordan smell strongly of oil. The British oil monopolies had an interest in the murder of the Syrian dictator Husni Za'im . . . (Landis, 1973, pp. 56–66)[1]

An Aramco employee, seeing the Senior Staff Camp and its handful of qualified Saudi inhabitants surrounded by Americans, might well believe anything possible of a company so large and so obviously non-Arab.[2]

The scrapping of the concessions systems for management contracts will make it more difficult to portray the majors as symbols of exploitation, particularly as national oil companies will be far more politically visible than former concessionaires and national governments will increasingly be blamed for blunders. It will also be more difficult to insulate labour relations problems from domestic political life by channelling the workers' grievances against foreign-owned companies. The kind of radicalism which naturally arises in the working force of a giant organisation will turn against the indigenous power structures.

There are many implications of this withdrawal of the majors from the front line of the political and economic life of host countries. For one thing, host governments are going to find survival increasingly tough when they have the responsibility of seeing that the oil industry runs smoothly and will be attacked if it does not. Secondly, a bone of contention between host and parent governments will be removed and should lead to a less tense relationship between the oil-producing world and the West as the buffer-cum-complicating role of the majors is reduced. The major impact of this will be in the Middle East and it may well be that there will be improved US relations with this area at the expense of the Soviet Union resulting from Arab awareness that oil companies can be persuaded to give ground.

The development of the oil industry seems to have resulted in a good deal of inequality within the producing countries. This impressionistic judgement is substantiated by Hollis Chenery's standard work on world inequalities, *Redistribution with Growth*. Of the sixty-six countries for which some form of statistics are given to indicate the degree of social inequality, four have grown up around the oil industry: Iraq, Venezuela, Iran and Mexico. Of these, only Iran is not included in the group of countries with the greatest degree of inequality (about one-third of the sample), and it only just falls outside the group limits. There can be little doubt that other oil-producing countries like Saudi Arabia, Kuwait and Libya would also show high degrees of social inequity if statistics were available (Chenery *et al.*, 1974). But this social inequality and the development of 'dual economies' did not arise from a conscious decision taken by the oil industry. It is more the result of the speed with which this sector of the economy was developed and the generally poor linkage between it and other parts of the local economy.[3] At the same time there are signs that local elites were hypnotised by the kind of modernisation which the companies seem to represent – a form of development which stresses giant, capital-intensive projects and plays down alternative approaches of more smaller and more scattered projects likely to lead to more equitable social development. The national plans of such elites give much emphasis on military hardware, steel, bauxite and petrochemicals and very little on the development of the countryside. If Third World governments emulate the type of industrial activities in which these companies have been so strong, they will perpetuate the kind of dual economic development of which the majors have for so long been a part.

BUFFER ROLE

One key point about the majors is that they have been wealthy non-governmental bodies whose existence between producer and consumer governments has added a degree of variety to international political relationships. Sometimes they have been a complication, making difficulties in areas which might otherwise have been quite harmonious. On other occasions, they have provided one or another of the governments involved with a means of realising national interests.

Companies as agents

Parent governments relied on the companies more or less officially to keep the oil flowing in the Second World War, the 1956 Suez

War and the 1973 embargo. More recent diplomacy around the International Energy Agency suggests that an increasing number of industrialised governments are willing to accept the companies, within certain specified limits, as agents. This role, of course, includes more than supplying oil as it involves, traditionally, a company-government information exchange. Often this is no more than swapping gossip but there are cases, such as with CASOC in Saudi Arabia, where companies acted as a diplomatic bridgehead when parent governments did not have diplomatic representation or where governments were on bad terms. Within two years of the 1956 Suez War, Shell was operating amicably in Egypt, while the British government was unrecognised. The same phenomenon was found in the case of France and Algeria where, during the war of independence, the French companies were one of the last links between the two sides.

The majors can also act as agents for their hosts. They carried out Arab instructions in the 1973 embargo and generally ensured that Arab oil did not find its way to Israel through their hands. They have carried out embargoes at other times, as when the Venezuelans prohibited the shipment of oil to the Dominican Republic in 1960. They are apt to become more clearly agents of host governments as management contracts increasingly tie them down to specific performance targets. However, there is a difference between a contractual arrangement and agreement by a company to work for a host government because of similar ideological commitments. Quite clearly the traditional majors are a product of the industrialised West although the international complexity of their activities raises the question of whether they do owe primary allegiance to any one country, even their titular parent. Many industrialised countries now seem to believe that the answer to this question is in the negative and have therefore created national champions.

Companies as instigators

If governments can get companies to act on their behalf, it is theoretically possible for the companies to get government help in their turn. As we have seen, such assistance has been rare. In the case of a *bona fide* dispute with a host government, the companies have generally – but not always – been able to count on some support from their parent authorities, generally some form of diplomatic representations or manipulation of aid-flows.

Companies as transmission belts

One of the most interesting roles the companies have fulfilled is that of a 'transmission belt' which receives and transmits messages between host and parent governments or achieves government policies. The foreign tax credit device was a policy put into effect by the companies which allowed Saudi Arabia and other oil-producing states to earn extra income. The decision to encourage other majors in Iran after the Anglo-Iranian dispute was a means of ensuring that sufficient Iranian oil would be marketed to enable the post-Mussadiq regime to survive. Likewise, King Faisal used the companies as intermediaries to pass on to Washington Arab reactions to US policies regarding Israel. Libya's expropriation of BP was a gesture of protest against Britain's apparent collaboration with Iran in the Gulf.

One aspect of this transmission of messages through the companies is that tax or regulatory decisions in one country can have ripple effects throughout the system. The foreign tax credit policy encouraged the overseas activities of US companies, but the later decision to rethink what kinds of foreign payments qualified for this credit played havoc with the Indonesian production-sharing contracts, and opened a large area of uncertainty in the negotiations over the Aramco takeover. At least part of Anglo-Iranian's problem with Iran in the late 1940s was that the British had imposed limitations on dividends which severely restricted the payments the company could make under the revised 1933 concession agreement (Elwell-Sutton, 1955, p. 81; Monroe, 1963, p. 111). Should any of the divestiture proposals pass into US law, then the implications for the rest of the world will be far-reaching, just as the closing of US markets to unrestricted imports in 1959 ensured that Europe and Japan would be the scene of fierce marketing competition by international companies denied one of their more interesting markets.

Companies as a complicating factor

There are the occasions when the companies have just been a nuisance to parent governments wishing to maintain good relations with a host country. A notable case was when the restrictive aspects of the formula determining the lifting of Iranian crude oil became a matter of extreme concern to the Shah in the mid-1960s. There was also the case of Anglo-Iranian's dispute with Iran in the early 1950s when a number of American and British officials were convinced that the company's intransigence was complicating the search for an acceptable solution. In Latin America, the State Department had to pressure Creole more than once to drop an intransigent line towards Venezuela (Wilkins, 1974, p. 221; Tugwell, 1975, p. 52). There is no doubt that the companies' existence imposed constraints on parent governments. The British have been hampered in drawing up a policy for North Sea development by the fact that OPEC might have taken extreme nationalistic measures there as precedents to use against Shell and BP elsewhere. Later they were constrained by the fear that the USA would discriminate against BP in Alaska should American companies not be given a share of North Sea spoils. The existence of other American companies in Peru meant that US reactions to attacks on IPC could have resulted in retaliations. There has been no other industry like oil for touching off extreme and emotional reactions. All the major expropriations, such as Mexico in 1938 or Iran in 1951, have triggered serious crises which have adversely affected the relationships of parent government and host. In the early days of Castro's rule in Cuba, reaction of the oil industry led to expropriations which were one of the main turning points in the US-Cuban relationship.

ECONOMIC ROLE

It is difficult to show that the majors have played a political role of great importance. These companies are primarily economic institutions, and this fact sets them apart from nation states, to whom economic gains and losses are but one set of items on their balance sheet. The point about economic actors is that so long as they can function reasonably well, they generally accept the political status quo. No company can afford to turn its back on profitable markets for the sake of a political principle. Nor can any resource-based company, like the oil majors, turn down the chance of developing important new deposits. Of course companies have to pick and choose between possible ventures, for none has the financial and managerial resources to take every opportunity that comes along. But within these limitations, they are automatically drawn toward the potentially most profitable and risk-free options open. The political climate of the countries in which these ventures fall is only one of the factors taken into account. The per capita income, the extent to which the competition is already established and the regulations governing things like levels of profit remissions are generally far more important than the political complexion of the guiding regime. What does matter, though, is the degree of stability that a government is perceived as having. Far better, for instance, in the eyes of many businessmen, to have a slice of an East European market, than one of a smallish LDC, run by an insecure dictator, even if he is pro-Western. The former may be less ideologically compatible, but the potential market is bigger and the communist world has a reputation for sticking to agreements and repaying debts.

The political tactics available to companies for gaining access to promising markets are limited. As the history of the oil industry suggests, it was probably only in the very earliest decades that companies could afford to behave arrogantly, confident that local governments would never evict them. However, as host governments grew in confidence, so the companies had to assess the impact of their tactics, and the hard-nosed approach was seen to be counterproductive, particularly as the industry moved away from long-term concession agreements to shorter, more competitive contractual arrangements.

In fact, oil company strategies are predictable. Obviously personalities have mattered on occasion, for one doubts if Shell would have withstood the challenge from Standard Oil and Jersey Standard if it had not been for Deterding's leadership. But the underlying economics of the industry make it possible to predict the general direction in which companies will move. Some will obviously move faster and more effectively than others, since managements differ in quality, but knowing who is leading a company at any moment is generally of little value in comparison to knowing its technological strengths and weaknesses and how these adapt to changes in the underlying economic and competitive trends.

It would be possible to simulate the development of the oil industry from the turn of the century by building into the model the location and ease of access of the various oil deposits; the sources of existing oil production; the size, development and location of the world's leading economies; some facts about the motivation of the imperial powers; some assumptions about the behaviour of companies in an international oligopoly; and some information about the level of government experience in most of the potential producing countries. Given these conditions, it would be fairly logical that the structure of the industry would have developed much along the lines of the way it has actually done. The sheer size of the US market, and the fact that there was a significant oil industry already in existence in the USA meant that American oil companies were bound to play a dominant role. The imperial rivalry of the Great Powers before 1914 meant that they were each bound to encourage the search for oil by national champions. The pre-eminent position of the British and their hold over Persia and its deposits meant that a significant British presence in the industry would probably come about. The two periods of US expansion into the Middle East can be explained by the dynamics of the US industry, scares about oil shortages, and the fact that once the scale of Middle Eastern reserves became apparent by the early 1940s, no self-respecting American company could afford not to get involved. One might also assume that the existence of oil production would bring about a reaction from host governments, which would inevitably become more sophisticated as they acquired wealth and some of their citizens got practical experience. Obviously one could not predict that it would be the British company in Iran which would have the most serious confrontation with a host government, rather than, say, an American company in Venezuela, but the trend toward greater host government control of the industry would seem inevitable.

Compare this with the problems of simulating the development of events which are less overtly concerned with international economic transactions. Take, for instance, Germany's relations with its European neighbours after 1918. Obviously all analysts would agree that Germany's defeat in the world war and the terms imposed on it by the peace treaty would inevitably lead to political parties demanding the restoration of the country's former central role in Europe. However, it is another matter arguing that there was an inevitability in the rise of a Hitler, who was both able to master the German political process to win supreme power and, at the same time, was unable to judge the exact point at which to stop his foreign adventures. Premature death, variations in the personalities of key actors, slightly stronger responses from neighbouring powers in the early part of his career as Chancellor could all have produced a qualitatively different history for Germany and, for that matter, the world. And there are any number of other examples one can give showing the complexity of non-economic issues compared with economic ones. Who would care to predict, for instance, the future of Sino-Soviet relations in the light of Mao's death? Where was the inevitability about the creation of the State of Israel? Although there are economic forces behind the proliferation of nuclear power plants, who can predict if we will be successful in containing the related threat of the proliferation of nuclear arms industries, before some region, or perhaps the whole of the world, gets obliterated?

These are life and death issues, and understanding the oil majors gives one little insight into them. Certainly, oil diplomacy is a valid subject of interest in its own right, and there are parts of the history of the oil industry about which one would like to know a lot more. A good political biography of some of the early industry leaders, such as Deterding, would be interesting because the oil diplomacy of the 1910s and 1920s was exceptionally complex, and a study like this would throw more light on the extent to which such characters were political forces to be reckoned with in parent government policies. Again, the memoirs of someone like Howard Page, who went through so many oil battles in the Middle East during the 1950s and 1960s, would undoubtedly be fascinating and throw intriguing light on this particular period of oil diplomacy.

One suspects, though, that opening the archives of an Exxon or a Shell would only add a limited amount of richness to the analysis of international relations. Before the 1970s, the majors just did not play a particularly interesting role in non-oil matters, and the fact that oil only sporadically caught the attention of diplomats meant that the majors did not often come close to the centre of the diplomatic stage. Certainly their archives would increase our understanding of the domestic political process of some countries and might illuminate certain bilateral relationships such as that between the USA and Saudi Arabia, but, in general, the fate of such oil-linked relationships has been quite low down the agendas of the Great Powers of the early to mid-twentieth century.

Paradoxically, as oil has become more crucial as an issue area, the independence of action that the companies have had within their chosen economic sphere has been eroded at both ends of the industry. The second half of this book examines in some depth how the majors have increasingly been relegated to the side-lines, and, although we want to know far more about oil diplomacy, an understanding of company politics has become fairly irrelevant.

This is why, when the general histories of twentieth-century international relations come to be written, the majors will rarely be given more than the occasional footnote on the pages dealing with the oil crises of the 1970s. This is not to argue that the majors are going to fade away as industrial bodies; nor that the technological and commercial strategies of such companies will not be of intense interest to international economists; nor that an understanding of these economic forces is not a precondition to a serious analysis of oil diplomacy, since they set the boundaries within which diplomats are constrained. But, though oil and energy issues will remain central to world politics in the coming decades, it is the diplomats who will lead the debate. Oil companies – even those the size of the majors – will merely be interested bystanders.

NOTES

1 See also Hirst, 1966, for analysis of the oil industry's impact on Arab public opinion.
2 Cheney gives a contemporary American view of the tensions at work among educated Saudi Arabians (1958).
3 For the case of Anglo-Iranian see Fesheraki (1966, pp. 20–1).

7
Hosts and Companies in the 1970s

The years since 1969 have seen the third redefinition of the power relationship underlying the oil industry. Average host government revenue per barrel rose from $0·91 in January 1970 to $10·98 in October 1975 (Shell 1975, p. 14). Consumer governments, which once treated their host counterparts with disdain, went firmly on the defensive as the Third World sought to restructure the world economic order under OPEC's guidance, and the international oil companies, which once had the contractual freedom to produce and market oil from their concessions as they saw fit, have either had to accept participation by host governments, or, in some cases, accept expropriation. The 1973 oil embargo showed that the Arab producer governments saw oil in an overtly political light, and the fact that the companies apparently obeyed Arab instructions raised the suspicion in some Western minds that the majors, far from being agents of the consumer nations, should now be seen primarily as agents of the hosts. Company-government relationships in the 1970s bore little relationship with those of the previous decades.

WHAT HAVE THE COMPANIES LOST?

Most visibly, the majors have lost the freedom to orchestrate OPEC's crude oil production.[1] In particular, they lost their freedom to react to developments in the market place without reference to the wishes of the producer governments who had suffered from the fact that the posted price of Middle Eastern oil had fallen in real terms every year from 1947, with the exception of two or three years in the mid-1950s (Adelman, 1972, chs 5 and 6; Blair, 1976, p. 117). OPEC's creation had served notice that the majors could no longer cut posted prices unilaterally, but it was not until the 1971 Teheran-Tripoli negotiations that host governments confirmed that they had the power to force an increase in posted prices against the wishes of the companies. When OPEC's six Gulf states decided in October 1973 to negotiate no further with the companies and to post their own prices for crude oil, the era when it was the majors which took most of the key pricing decisions in the oil world had ended.

At the same time, the companies were forced to accept the abandonment of traditional concessional arrangements in favour of host government participation in, or outright expropriation of, their upstream activities. It is difficult to make clear distinctions between these two issue areas (price and concessional arrangements) since negotiations about the latter have increasingly involved discussion of the price discounts to be granted to traditional concession holders in return for various services which they are still willing to provide. But it is still important to spell out what the oil companies have conceded in recent years and what their newer relationship with host governments seems to be.

End of the concession system

Events during the 1970s were the culmination of a series of attacks on the traditional system of concessions which dominated the oil industry during the earlier years of this century and, though occasionally modified at producer government insistence, still reflected the original principles of the system. Typically the concessionaire was granted exploration rights over extensive tracts of land for a period of fifty years or more, with no real host governmental control over what they did there. The Venezuelans entered the 1970s with the bulk of their concessions not due for reversion until 1981, while Aramco was allegedly safe in Saudi Arabia until 1999, the Mosul Petroleum Company in Iraq until 2007, and KOC in Kuwait until 2026 (Church Report, 1975, p. 85; Smith and Wells, 1975, p. 566).

Such agreements generally did not give the host government any say in the planning or running of activities in the concessions. Most did not specify commitments on the part of the concessionaires to spend given amounts of money on exploration within given times, nor production rates once oil or gas was found, nor reinvestment ratios for profits which might be made, nor the development of, and support for, local suppliers. In other words, apart from the freedom the host governments had at the start to choose the original concession holders from a limited number of suitors, they had no rights to a strategic role in the development of the resources in question, once the choice was made. When the Shah of Iran asked the Iranian Consortium to increase production in the late 1960s, he was in the position of being a powerful suppliant but his legal right to make such a demand was in doubt. One part of the changed environment of the 1970s is that the companies have increasingly been forced to accept that host governments should have the right to make strategic decisions at the upstream level. This is of course a genuine revolution at the formal level. Where one can be more sceptical, though, is at the functional level where it is legitimate to point out that host governments are still faced with the problem of coming to terms with companies which still dominate certain key processes in the oil industry.

There is no pretence that the majors will be totally ejected from the economies of conservative host governments. The Saudis, for instance, have moved with agonising slowness in seeking 100 per cent participation in Aramco's activities since 1969 when Sheikh Yamani first openly started talking about participation as an alternative to outright nationalisation. ('Clearly, nationalisation would be a disaster for all concerned in oil affairs.') (Madelin, 1975, p. 74; Penrose, 1975, p. 40.) But the actions of more militant hosts like Libya, Algeria or Kuwait meant that the original demands which the Saudis got OPEC to accept in 1972 for 25 per cent government participation, rising to 51 per cent control in 1983, appeared too timid and in need of strengthening. Despite this, the Saudis have sought to delay any final deal for 100 per cent participation in Aramco, apparently seeking a compromise which will look sufficiently nationalistic on the surface to protect the government from the charge of having failed to promote the national interest, while, at the same time, being sufficiently gentle to leave the companies as partners in the further development of Saudi oil production and the general industrialisation of that country (Stobaugh, 1975, p. 179).

In the Venezuelan case there has been no figure similar to Yamani able to steer the course of a generally moderate policy. Oil has been a burning political issue in Venezuelan politics from the 1920s and a leading political figure, Perez Alfonso, has opposed any involvement of private capital in the oil industry be it foreign or domestic (Tugwell, 1975, p. 175). There has been very bitter public debate about the role that the foreign oil companies should be allowed to play in the post-nationalisation industry. Despite all this, the majors are still getting privileged access to Venezuelan crude oil, thanks to the device of service contracts whereby the companies sell some form of expertise to the government or its agencies for a fee which may well take the form of a reduction of some 10 or 20 cents in the price of each barrel of crude oil which it purchases. Exxon's Venezuelan subsidiary was nationalised on 1 January 1976 and now operates as a government company, Lagoven, but there is an agreement with Exxon Services Co. Inc. to provide technical assistance to Lagoven in return for fees based on Lagoven's production of crude oil and natural gas liquids and refinery output. Exxon 'lends' its ex-subsidiary some 150 employees while the Venezuelan government's programme to train some 10,000 engineers is realised. Nationalisation appears less impressive if the former company still supplies key personnel, the most important markets and perhaps a certain amount of strategic thinking (*Oil and Gas Journal*, 12 January 1976, pp. 36, 51).

Host governments will obviously seek to keep their dependence on the old concessionaires to a minimum. Kuwait, which has relatively few problems in crude oil production, has signed oil-supply agreements with Gulf and BP in which the latter guarantee to take, and are guaranteed the supply of, a given range of oil during the duration of the contracts at a discount of 15 cents a barrel. These are the primary deals into which Kuwait has entered, and though there are less significant service agreements for the majors to supply personnel, there are no long-term commitments. If the Kuwaitis decide they do not need such help, then the companies will not provide personnel. Either way, the companies' access to crude oil supplies is not affected (*Petroleum Intelligence Weekly*, 29 March 1976, p. 4; 24 May 1976, p. 1).

Host governments are thus entering into 'agency contracts' or even joint ventures (though these are still quite rare) whereby the international oil companies offer some combination of markets, technology and skilled personnel in return for some combination of fees or preferentially priced crude oil. What is important to note is that the vast majority of concessionaires have been offered some form of medium-term supply contracts, even if the amount of crude oil made available is less than what they would have had under their former arrangements, and even if the value of the discounts are eroded by price-shaving by competing national oil companies (Smith and Wells, 1975, p. 582). It is rare for the companies to have any equity stake in the new producing entities (thus underlining the distance the oil industry has moved away from the old concession agreements), but there are circumstances in which companies can continue to exercise significant influence on key decisions under the new relatively unstructured conditions. The most telling factors are not surface indicators such as movements in equity stakes, but more nebulous ones showing the exact balance of influence which is emerging case-by-case between companies and local policy makers. These vary depending on the sophistication of the country concerned and the complexity of the technical problems posed by the oil industry in question, but there will undoubtedly be individual cases of companies which, despite losing formal control, will still have *de facto* authority over decisions which are often of some significance for the management of oil producing. This authority would be expressed in decisions involving the choice of technologies, markets and perhaps even the viability of moderately ambitious projects. However, one of the consequences of events since 1969 is that host governments or their agencies now want to take, or at least to be seen to take, the key strategic decisions themselves. The question for the late 1970s is the extent to which they can take these without reference to the view of the international companies, and the extent to which decisions, once taken, can be implemented in the face of potential corporate withdrawal.

Increasing producer government 'take'

Host governments were forced to rely primarily on a fixed payment (royalty) per barrel of crude oil produced until an effective demand for some form of profit-related tax emerged. During the 1940s Venezuela put forward the fifty-fifty formula by which companies paid to host governments half the difference between their allowable costs and the 'posted price' for crude oil which was then close to the market price. By its very simplicity, this system swept round the oil world and, though the host governments would on occasion get more than 50 per cent of these notional profits, the principle was that their take could be calculated from the posted prices declared by the majors. It was inevitable, therefore, that posted prices became charged with a political significance they had not hitherto possessed. This became clear in 1959 and 1960 when the majors lowered the posted price in response to price cutting in the market price and triggered a wave of protests which culminated in the creation of OPEC. From then on it was clear that a ratchet effect was operating; whatever happened to world markets, the posted price was not to be reduced. It thus became a tax-reference device, divorced from market forces, and as Adelman commented, '1960 marks a long step away from income taxes and back to per-barrel payments, which governments have gradually raised' (1972, p. 207).

For a decade after its formation, OPEC and the majors tested each other and generally came out about even. OPEC was unable to get the companies to restore the 1960 cut in the posted price, but avoided any further cuts which were conceivably justified by market conditions, and even won a slight increase in government take (which went from 80 cents a barrel in 1960 to 95 cents in 1970) by insisting that royalty payments should not be set against income tax. The Teheran-Tripoli agreements were the first significant battle that the majors lost over the revenues to be paid to host governments. Although the fifty-fifty deals of the late 1940s and early 1950s also gave the host governments a considerable increase in income, most majors were not too worried about conceding the fifty-fifty principle, once it was established that they could offset their upstream tax payments against their downstream ones. Teheran-Tripoli was different. This time there were no tax angles to protect the majors and higher payments to host governments could only be recouped providing the market would permit end-prices to be raised proportionally. Should this be impossible, then increased payments could only come from decreased indirect taxation by consumer governments (unlikely), or from company profits. This marked the crucial difference from the fifty-fifty arrangement. The Teheran-Tripoli Agreements were therefore a watershed in the history of the industry. Until 1971, the companies determined the oil revenues of host governments by unilateral decisions to adjust the posted price in conjunction with their general ability to balance production rates between various sources. They did not worry much about hints of change during the 1960s such as the creation of OPEC in direct response to the unilateral cut in the posted price in 1959, or the growing determination of Iran to demand faster increases in crude oil production or payments in advance. One of the constants of the industry had seemed to be that the companies could fix posted prices without any significant reference to the interests of the host governments. Teheran-Tripoli proved conventional wisdom wrong. The companies – the independents and then the majors – were forced to negotiate the levels of posted prices with host governments and the latter began to sense that the power of the companies was less impressive than they had feared.

In retrospect, the increases the companies actually conceded were modest, letting government 'take' rise from $0·90 per barrel of market crude oil in January 1970 to $1·20 in March 1971 and, after subsequent negotiations on issues like compensation for the falling dollar, to $1·80 in March 1973. Then, just as the collapse of a dam is often forecast by minor leaks in its structure, the traditional defences of the oil companies against the demands of host governments collapsed. The latter's take rose sixfold in the next two and half years and reached $11·17 in October 1975 – a figure set not by the companies, but by OPEC. This massive increase was, to some extent, the result of special circumstances. In particular, the Arab-Israeli War enraged Arab producers to the point of declaring an embargo against Western powers such as the USA in the context of a general restriction of oil production aimed at putting pressure as well on governments which were exempted from the formal embargo. The combination of these two tactics took 5 million b/d off the world market or 9 per cent of non-communist production (Stobaugh, 1975, p. 181). Hardly surprisingly, there was some panic on the part of consumer governments and companies which played into OPEC's hands. It is probable, as I have argued elsewhere, that the degree of panic ascribed to governments, particularly European, has been exaggerated, but there is little doubt that there existed genuine consternation among the many smaller companies which had been relying either on the Rotterdam market or on more or less formalised supply arrangements with various crude-surplus majors. As the latter found they had to concentrate on supplying their own affiliates and those outside clients with whom they had binding contracts, these companies with marginal arrangements were faced with the possibility of being without any crude oil at all.

The result was that they bid enthusiastically for oil which Iran, Nigeria and Libya put up for public auction, offering prices from $16 to $20 a barrel. These bids were then used by OPEC ministers to justify major raises in the posted prices for crude oil (Penrose, 1975, p. 51). The majors found themselves unable to resist OPEC price setting and it was not until the winter of 1975–6 that the companies were able to show that they could once again influence, if not control, the prices at which they purchased oil from host states.

Pricing – a marginal role for the majors?

Circumstances after the 1973 embargo were such that no major had any practical alternative but to accept OPEC decisions on the levels of posted prices. Political tempers were running high enough in the Arab world for an unco-operative company to run the risk of being expropriated like American and Dutch interests in Iraq after the outbreak of the Yom Kippur War. The embargo produced such a tight fit between supply and demand that the contributions of no one company were so essential that they could not be replaced by some competitor. Moreover, the companies had grown used to dictating the levels of host government take, and it was to be a while before they adjusted to the reverse situation. Furthermore, the whole pricing issue was inextricably entwined with that of participation and the diplomacy surrounding this was of considerable complexity in its own right. The post-1973 period was one in which the bulk of the host oil-producing states were moving from a situation in which they had around 25 per cent of the operations of the former concession holders, to some 60 per cent during 1974, and then to something approaching full control from the winter of 1975–6 onwards. While all companies were concerned about the increase in the absolute levels of the price of oil, the former concession holders were able to 'buy' price-breaks on 'equity crude', that is to say on the proportion of the crude offtake equivalent to their equity stake in the former oil-producing companies and consortia. For most of 1974 and 1975 the majors were interested in resolving the participation issue to their advantage and thus did not want to antagonise host governments by trying to drive down the price of oil. Until participation was resolved, the majors were more concerned with attractive prices for their future crude oil supplies than with prices for immediate customers. But once the principles for setting differentials had emerged, the majors were in a position to turn their attention to the absolute level of prices and found that they still had some marginal leverage over host producing governments in determining pricing policies.

The main factor which started to work in their favour was that supply and demand of crude oil had responded to the large price increases imposed by OPEC and there was, as a consequence, an excess of potential productive capacity. Demand constraint was already noticeable in the summer of 1974 after the relaxation of the embargo (*Petroleum Economist*, July 1974, pp. 243–5). World demand for OPEC oil, which had climbed steadily from 1960 to 1973, fell just over 1 per cent during 1974 and 11·4 per cent during 1975, and a production level resulted which was 27 per cent below estimated operable productive capacity (*Petroleum Intelligence Weekly*, 2 February 1976, p. 7).

This decline in oil demand reflected the world slump which started during 1974, conservation policies and the substitution of other sources of energy for oil. It meant that the companies for the first time since 1969–70 could meet their market needs and yet refuse to take crude oil from unattractive sources such as host producer governments whose prices got out of line. Though this device of refusal was first aimed at relatively minor producers, it was, by 1976, apparently capable of being used against everyone but the Saudis, and perhaps the Iranians.

The market forces which are to the advantage of the companies stem from the fact that the differences in the ultimate market values of different crude oils vary greatly and for different reasons. One immediate impact of the fall in oil demand for instance was a collapse in the cost of freight haulage, thus making Middle Eastern oil suddenly much more attractive in price relative to crude oils from short-haul destinations like Libya and Algeria which, up to 1973, had been pricing their output at a premium compared with, say, Iranian crude oil. In addition, the shock of the embargo temporarily lessened the general enthusiasm of the industrialised world toward the environmentalist movement, with the result that the emphasis on the use of low-sulphur crude oils declined. This meant that a range of producers, including Libya, Algeria and Abu Dhabi, with pre-embargo price structures which valued low-sulphur crude oils at a premium, increasingly had difficulty throughout 1974 in getting companies, independent refiners as well as majors, to pay the extra premium at which they were selling low-sulphur crude oil. Finally, each of the products ultimately derived from oil was affected differently by the general price increases. By 1976 it was clear that demand for the lighter end of the barrel (products like petrol) had been less affected than that for the heavier end (e.g. the fuel oils used in industry). This differential impact reflected the fact that there was no ready alternative to petrol for powering cars and lorries, while industrial users and utilities could at least partially substitute coal or natural gas. Since crude oils vary in their composition, a change in the market demand for final products was reflected in a change in demand for different crude oils. Iran, for example, was increasingly hard put to find customers for its relatively heavy crude oil except at a discount.

The growing resistance of purchasers showed itself in extreme fluctuations in oil-flows from certain of the countries whose price structures had got out of linc for the kind of reasons suggested above. By February 1975, for instance, Abu Dhabi's exports of crude oil were only 39 per cent of what they were in July 1974 and it was forced to cut the posted prices of premium crude oil. At the same time, Libyan exports had fallen an equivalent amount from January 1974, and Libya too was forced to bring its prices into a more realistic relationship with competing crude oils. Algeria faced the same pressures, but was rather faster in adjusting its pricing policies so that exports did not fluctuate to the extent of those from Libya and Abu Dhabi.

IN THE MIDDLE: COMPANIES AND THE OIL EMBARGO

The embargo symbolised the way the power balance within the industry had changed. Not only had the host oil producers learned how to improve their financial returns, but they did not feel vulnerable to the kind of pressures imposed on Iran during the 1951–4 crisis. They realised that they were in a position to move beyond economic to political goals. The majors, Aramco's partners in particular, were well aware of the pressures building up in the Arab world and Aramco's standard briefing papers in mid-1973 were specifically designed to warn visiting American dignitaries of the problems being caused within Saudi Arabia by US support for 'the expansionist policies of the current Israeli Government' (Church Report, 1974, pt 7, p. 525). It was implied that similar messages had been given to a string of visitors dating at least as far back as 1968. The dilemma posed for Aramco existed even before the creation of Israel in 1948 but Aramco could live with the tension caused by the conflicting aspirations of its host and parent governments until 1973. Then, however, the Saudis, who were being pressed by Egypt into backing a confrontationist policy toward Israel, began to put much more specific pressure on the company, pointing out that its concession could be at stake should the Saudi regime become isolated within an Arab world which was becoming ever more belligerent in the face of US Middle Eastern policies.

The response of Aramco and its partners was to launch an unprecedented campaign in the USA stressing the need for a more balanced policy toward the Arab-Israeli conflict. This was far more than the usual run of interaction between multinationals and a foreign policy establishment whereby the two sides swop views on political developments in commercially important parts of the world and the businessmen more or less discreetly inquire about possible modifications to policies or regulations which particularly affect them. In this case, the relevant companies sought to change American policy and were even willing to appeal over the head of the State Department to public opinion. The Saudi-connected majors felt they had a responsibility for getting the views of their host government across to the US administration. It is particularly interesting to note that at least one key member of the State Department, Jim Akins, was willing to use the companies as means of circuitously influencing his political masters. Although the Aramco directors grew steadily more worried about the situation, they do not seem to have felt it necessary to take significant action until mid-1973. They were disturbed by a meeting with King Faisal on 23 May at which he spelled out the growing danger of Saudi isolation in the Arab world. Five directors then promptly made the rounds of the State Department, White House and the Department of Defense, summarising the drift of Saudi oil policies over the previous year, and stressing Faisal's points including the threat that all American interests would suffer if US policies toward Israel remained unchanged. 'Action', they quoted Faisal as having said, 'must be taken urgently; otherwise everything will be lost' (Church Report, 1974, pt 7, p. 509). The oil men were met with disbelief, by a feeling that Faisal was crying wolf, and by an apparent conviction that there was little the USA could do, even if it was willing, to affect the Arab-Israeli issue in the immediate future.

The Aramco partners did not take this rebuff as final. Mobil's president, William Tavoulareas, directly lobbied the Assistant Secretary of State, Joseph Sisco, and his company took an advertisement in the *New York Times* in June to argue the need for a more balanced line toward the Arabs.[2] In July Socal's chairman, Otto Miller, sent a letter to his stockholders. However, such expressions of oil company thinking merely seem to have stirred Zionist protest and made no perceptible impact on US policy makers. The chief ally of the oil companies during this period was James Akins, former head of the State Department's Office of Fuels and Energy (1968–72), author of the influential 1973 article in *Foreign Affairs*, 'The oil crisis: this time the wolf is here', and Ambassador to Saudi Arabia from 1973 to 1976. His relationship with Aramco was an extreme instance of those rarely publicised alliances which sometimes occur between companies and parts of the foreign policy establishment. In January 1973, while assigned to the White House, Akins contacted Aramco's Washington lobbyist, Mike Ameen, to inform him that the President's chief domestic aide, John Erlichman, was shortly to visit Saudi Arabia and hoped to meet Yamani. Akins asked Ameen, who was leaving the next day for Saudi Arabia, to tell Yamani that it was particularly important that he 'take Erlichman under his wing and see to it that Erlichman was given the message: we Saudis love you people but your American policy is hurting us'. Subsequent to the Arab-Israeli cease-fire there was considerable American optimism that the Arab oil embargo on the USA would be lifted. Akins, then Ambassador to Saudi Arabia, was more pessimistic and sent a message via a local Aramco manager that industry leaders back in the USA should use their very best contacts in the Nixon administration 'to hammer home [the] point that oil restrictions are not going to be lifted unless [the] political struggle is settled in [a] manner satisfactory to [the] Arabs'. He urged these executives to put their case as bluntly as possible, pointing out that part of the industry's problems with the American government stemmed from the 'industry presentation not being direct and to the point' (Church Report, 1974, pt 7, pp. 423, 517, 546–7).

However Aramco and its partners were keeping up a steady flow of advice to US officials throughout this eventful autumn quite independently of what Akins might be advising his State Department masters. Six days after the start of the Yom Kippur War, the four Aramco partners sent President Nixon a joint memorandum warning that any increased military aid to Israel could trigger retaliatory Arab action against oil-flows which, in a situation where the world oil industry was already fully stretched, could prove disastrous to US interests in the Middle East as Japan and Western Europe would be lured into expanding their supply positions in this area. At the same time they referred to the key negotiations between OPEC and the oil industry which were being carried on in Vienna and could lead to serious balance of payments problems for the Western world (Sampson, 1975, p. 252). However, by the time General Haig passed this message on, the key decision to authorise the airlift of supplies to Israel had already taken place (Church Report, 1975, pp. 145–6). On 15 October Aramco chairman Frank Jungers passed on a warning from Yamani that the USA could well face an oil boycott. Two days later the Saudi Foreign Minister, Omar Saqqaf, handed Nixon a confirmatory letter from Faisal stating that if the USA did not stop its supplies to Israel within two days, there would be an oil embargo. By that time, the die was cast.

Up to that moment, there is a case for arguing that Aramco and partners were functioning more as agents of the Arab hosts than of their parent governments. With the implementation of the embargo, we move into a period in which the true ambiguity of the oil companies' role in today's world becomes starkly apparent. The companies implemented the instructions given them by the Arab governments. When they were told to cut production, they did so without quibble. When Aramco was told to cut production by 10 per cent and then, on top of that, shut in liftings equivalent to that which had been produced for ultimate sale to the USA in pre-embargo months, the company dutifully cut production back 23 per cent below September levels. When the Saudis insisted that Aramco impose tight control over the destination of its oil, the company got tanker captains to sign affidavits as to their destination and arranged to receive cabled acknowledgements of each ship's eventual arrival at the approved port or terminal (FEA, 1975b, pp. 2, 21). The result was that the letter of Arab instructions seems to have been followed quite closely. Imports of Arab oil into the USA dropped from 1·2 million b/d in August–September, to a mere 19,000 b/d, and that seems to have resulted from the belated arrival of a tanker which had been loaded in Saudi Arabia before the crisis. A similar picture emerged for the Netherlands which was the other prime target for the Arab embargo (Sampson, 1975, p. 255).

Probably the most controversial incident during this period was the Saudi request to Aramco for details of their crude oil supplies to US military bases throughout the world. Mobil and Texaco were somewhat reluctant to release this information, but Exxon, after consultation with the Defense Department, felt that the Saudi request should be met, and the details were in fact provided. Given the fact that this occurred when international tension was at a peak, this 'leak' was awkward to say the least, and was the one incident above all others which was used in Washington to substantiate doubts about the patriotism of the oil companies (Church Report, 1975, p. 139).

However, the companies did not forget their customers. While observing the letter of the embargo, they effectively softened the effect, despite the lack of any consistent political guidance from consumer governments, by spreading shortfalls evenly around. There were differences in tactics between companies; some allocated supplies in relation to historic deliveries and others were more influenced by current and projected demand. Aramco appeared peculiarly sensitive to the political and economic demands of their host government. This was hardly surprising given the growing involvement of King Faisal in the Arab-Israel issue and the central and critically powerful position of Saudi Arabia in world oil affairs. In particular, the crude-short partner, Mobil, was very vulnerable to any loss of Saudi crude oil, and the other partners were wary lest Mobil undermine the overall Aramco bargaining position.

Dependence on hosts

Other companies, not tied to Saudi reserves, took a line which was less openly concerned with the Arab cause and more allied to the traditional concerns of the consumers. Gulf was unhappy with Aramco's equivocation when the latter was forced to choose between all-out resistance to posted-price hikes and the prospect of losing their Saudi stake.

Ironically, however, in December 1975 Gulf itself was to settle for a Kuwaiti takeover of the remaining 40 per cent of the Kuwait

Oil Co., a joint venture with BP, despite the intervention of the White House, which was worried about the precedent set by 'too easy a surrender'. Just as Aramco had preferred not to antagonise the Saudis, so Gulf, at the end of 1975, valued a continued presence in the Kuwaiti oil industry over any last-ditch defence of a principle which, it felt, primarily concerned the oil importers (Wilkins, 1975, p. 168).

Of all the majors, Shell seems to have been the most consistently opposed to making concessions to hosts. It took the hardest initial line against the early Libyan demands and was forced to stop all production in September 1970. According to one of the participants, it was this company which first floated the idea of calling the majors and independents together in January 1971 to formulate a common position in the face of escalating demands from OPEC. In the subsequent negotiations held within the framework of the London Policy Group, in which all the relevant companies took part, it appears that Shell was noticeably less willing to accede to the Gulf countries' demands than others (Stobaugh, 1975, p. 184). In mid-1973, Geoffrey Chandler, a Shell executive, publicly stressed the need for importing nations to agree on sharing supplies in the event of a shortage, thus reflecting the concern then felt, at least within London oil circles, about the deadlock on this issue with the OECD.

In so far as any company was going to take such a relatively uncompromising line towards the producer governments, it was always likely to be Shell. Unlike most of the other majors, its international fate was not particularly tied to what happened to its productive facilities in any one country or geographical area. Also, the general diversity of its crude oil production meant that it stood to lose a great deal from the wider acceptance of concessions made in any single country. In addition, the fact that it has been a relatively crude-short company with a strong reputation for marketing oil has probably made it particularly aware of the problems which a significant increase in the price of oil would cause in importing countries.

The other major which had distinctive experiences during the early 1970s was the French company, CFP. Because of its share in the Iraqi and Iranian consortia it was a member of the London Policy Group during the critical months of 1971. But CFP had a range of interests which were sufficiently distinctive from those of the Anglo-Saxons to set it somewhat apart. Iraq was a case in point where, in the previous decade, it had tangled with fellow participants in IPC by obeying French government instructions and entering negotiations with the Iraqi government over the exploitation of a field which fell within the disputed territory expropriated in 1961. In 1972 the Iraqis nationalised IPC, but came to an agreement with CFP whereby the latter was allowed to take its traditional share of the nationalised oil. The rest of the IPC partners seem not to have been too worried about this apparent rank-breaking, since CFP was able to act as a discreet mediator on behalf of the others, and a settlement was in fact reached between the two sides in March 1973 though CFP did get favourable treatment as far as price was concerned. The majors benefited from a partner politically acceptable to the somewhat touchy Iraqi regime, though they were uneasy over CFP's favourable treatment thanks to a direct government-to-government agreement between France and Iraq in the immediate aftermath of the 1972 IPC expropriation, in which the French policy toward the Palestinians was specifically singled out for mention (Tanzer, 1969, pp. 75–6; Mosley, 1973, p. 338; *Petroleum Press Service*, 1972, pp. 238–9, 301, 327; 1973, pp. 107, 124–7). However, though CFP and ERAP could benefit from the political stance of their parent government, they could also lose. The terms for the exploitation of Algerian oil were fixed by intergovernmental agreements signed in 1962 and 1965 and when the Algerians opened the price issue in 1969, the negotiations were carried on a bilateral basis, not within the multilateral framework of the Teheran-Tripoli negotiations. However, the Franco-Algerian negotiations could not be insulated from the impact of the decisions made in these parallel talks. The Franco-Algerian dispute was more overtly political in nature, and the Algerians lost their patience and announced, in February 1971, that they were taking a 51 per cent stake in the French companies. The French, with the government taking the lead, decided to play the issue extremely hard. CFP and ERAP tried to emulate the tactics of the majors in Iran in the 1950s by suspending their crude oil offtake from Algeria and launching a global publicity campaign to scare off potential purchasers. This dispute spilled over into non-oil issues and the flow of Algerian products like wine was impeded. In the circumstances of 1971 US companies were only too happy to start negotiating for sulphur-free Algerian crude oil. CFP had to make its peace with Algiers in June, with the more intransigent ERAP hanging on to October (Aruri and Hevener, 1974, p. 79; Chevalier, 1975, pp. 76–81).

This is generally held to be a classic example of the danger of governments getting too closely involved in oil politics. Of course bilateral intergovernmental deals may pay off, as in the case of French diplomacy toward Iraq, but they are always potentially explosive. Once governments are involved, purely commercial disputes become inextricably entwined with non-economic issues and it is harder to keep negotiations at a low key. The French probably lost by not merging their Algerian negotiations into the wider multilateral diplomacy then taking place between the OPEC powers and the West in general. It is possible that CFP would anyway have been forced to concede relatively generous terms to Algeria, because of the latter's favourable geographical position and attractive crude oil. However, the fact that its Algerian arrangements had to be decided within a somewhat claustrophobic, post-colonial negotiating framework meant that the final compromise was inevitably a more bitter affair than the same kind of compromise which its Anglo-Saxon competitors were having to accept elsewhere in the world.

Host government agents?

There are those who argue that the companies have not fought at all seriously for consumer interests in the years since 1970 and that they have become agents for OPEC. As early as 1971, Walter Levy interpreted participation as a plot to bind company interests closer to those of the host governments until the majors became totally subservient. A year later, Morris Adelman accused them of being 'agents of a foreign power'. By 1976, the arguments had become more sophisticated and can be summed up by Anthony Sampson's testimony to the US Senate's Antitrust Committee:

> I . . . believe that it would have been very much more difficult, perhaps impossible for the OPEC countries to organize their cartel and maintain it so effectively if a few companies had not been dominant in the main producing countries, serving as the machinery for maintaining the OPEC cartel. And those companies now find themselves in a position of being closer in their interests to the producing countries than to the Western consumers. (1976, p. 4)

There is some supporting evidence for such claims. Did not, for instance, Eric Drake, the then chairman of BP, state in 1970 that the majors had become 'tax collecting agencies' for the OPEC countries? Were the companies not able to increase their profits from 1970 to 1974 at the very time when they were conceding higher takes to the host governments, thus suggesting that such concessions were not too painful? Also, has there not been the feeling that the majors did not always stand behind the independents in the struggle against higher tax payments with the enthusiasm they might have chosen to show? (Penrose, 1975, p. 42; Sampson, 1975, p. 234). Finally (the clinching argument for many), do not higher oil prices boost company profits and protect their investments in alternative sources of energy?

These allegations are not easy to counter and there is evidence that different companies have had different interests, thus leading to some disunity in the face of pressure from the host governments. The original breakthrough in the old price structure came in September 1970 when Occidental, a relative newcomer to the international scene, caved in to Libyan demands. However, this only occurred after Occidental and Exxon had failed to agree on terms by which the latter might offer the former replacement oil. Exxon insisted that such back-up oil should be at third-party market prices, which Occidental rejected and was thus forced to come to terms with Libya. This failure of a major and an independent to come to some form of mutually beneficial pact reflects a history of inter-company tensions in Libya whose oil was superbly placed for the European market. Unable to keep the independents out of Libya, the majors had undermined the favourable position the independents had won during the early 1960s by offering the Libyan regime better terms, provided that all companies were taxed in the same way (Church Report, 1975, pp. 121–5; Sampson, 1975, p. 216).

A different type of split within the industry showed itself in the 1972 discussions on participation between the companies and the Gulf states, in which the companies conceded an immediate 25 per cent stake to the host governments which was to rise to 51 per cent in 1983. The suspicion was aired that the Aramco partners had a much greater interest in such a deal than outsiders like Shell, BP or Gulf, since the planned expansion of oil production in Saudi Arabia was so great that Aramco could accept Saudi participation and still end up with an increase in the volume of crude oil which they could call their own. Other companies whose fortunes were tied to Kuwait or Iran had little scope for significantly increased production, so that any participation agreement would leave them with smaller volumes of crude oil. Gulf even went to the extreme of sending a cable to the team which was negotiating on the participation issue, arguing that their 'insistence on maintaining control of the oil for [their] own use has given strength to the OPEC position' (Church Report, 1975, p. 137).

However, such inter-company splits on policy were rare and the companies were generally agreed in trying to avoid conceding higher payments to host governments. They sought help from their parent governments in resisting OPEC demands and they specifically tried to centralise their negotiations with the host governments, so that weaker companies should not be picked off one at a time. Public divergence between the interests of different companies was certainly outweighed by instances of inter-company co-operation. Nevertheless, though the companies loudly protested, they still managed to 'have their cake and eat it', in the sense that their profits rose substantially even though they had given ground to the host governments. The charges against the companies can be summarised thus in the arguments of Peter Odell after reading an early draft of this chapter. Company profits stagnated through the 1960s and so, around 1968, a conscious decision was made to try upgrading profits. Unfortunately, the resultant attempt to get production in better line with global demand got out of hand as the OPEC countries not only got in on the act, but took the lead. This did not matter, since the companies benefited. As OPEC members raised the price levels for oil, the companies were able to pass these higher prices on and to maintain their proportional margins with the result that their overall profits rose. In particular, the companies gained from the higher world price of oil since it increased the profitability of their investments in high-cost, alternative energy sources such as North Sea and Alaskan oil, North American coal and nuclear power.

One piece of evidence supporting this argument that the companies may have conspired with the producers to raise oil prices is that the State Department's Jim Akins was arguing in the early 1970s that the price of oil should rise in gradual steps until it reached some $5 a barrel figure in 1980 (Akins, 1973, p. 479). It is certainly true that the majors had indeed been diversifying their interests both geographically and functionally through the 1960s. One indication of this geographic diversification is that the rate of discovery of crude oil reserves outside the Middle East, which had been static from the 1930s, started to rise in the late 1950s, to reach a level in the early 1970s (around 11 billion barrels a year) which was roughly double the rate found twenty years previously. In contrast, discovery of crude oil reserves in the Middle East, though running at a somewhat higher rate, reached a peak of around 22 million barrels a year in the late 1940s, and thence declined to a rate which was only some 4 billion barrels a year above that of non-Middle Eastern discoveries in the 1970s (Exxon press briefing). Some of these discoveries outside the Middle East were in countries like Nigeria, which were OPEC members, but some were also in politically 'safe' areas like the US and Canadian Arctic and offshore Europe and North America.

At the same time, the companies have been diversifying out of oil into other energy sources and, occasionally, into totally non-related fields (as when Mobil moved into the retailing industry through its acquisition of Marcor). Thus Conoco, Occidental, Gulf and Exxon all moved into coal; Gulf, Exxon, Conoco, Sohio, Kerr-McGee moved into uranium mining; Gulf and Shell moved into nuclear plant construction through their troubled General Atomic partnership; while Exxon moved into fuel fabrication. Their motivation seems to have been to reduce dependency on oil as its political future grew more uncertain, and as nuclear power increasingly seemed to be the energy source for the coming decades.

Moreover, the statistics of the 1970s give some initial support to the companies' critics. Company profits had stagnated from 1967 to mid-1972 with the overall profits of the Chase Manhattan's group of oil companies[3] rising a mere 45 per cent during that period. A spectacular boom in company profits started in the third quarter of 1972 which peaked in the second quarter of 1974, and the Chase group's profits increased roughly 150 per cent during this period. The fact that this boom was occurring even before OPEC initiated its major price increases of late 1973 and 1974 was an indication of the tightness of the world oil market which resulted from the short-lived world boom of 1972–3. Since the climactic events of 1973, the following picture has emerged of this Chase group:

Company profits: the combined net income (i.e. profits) of the Chase Group peaked at $16·4 billion in 1974, dropped to $11·5 billion in 1975 and rose to $13·3 billion in 1976 (provisional figures). These compare with $6·5 billion in 1970.

Return on companies' average invested capital: peaked at 19·2 per cent in 1974, fell to 12·8 per cent in 1975 and rose to 13·8 per cent in 1976. The figure for 1970 was 10·3 per cent.

Companies' net income as a per cent of total revenue: 6·7 per cent in 1974, 4·7 per cent in 1975, 4·8 per cent in 1976. These are record low figures, roughly half the rates found in 1965 and 1966. With the exception of 1973, this figure has been falling consistently.

Direct Taxes: $39·6 billion in 1974, $36·8 billion in 1975, $35·5 billion in 1976. These compare with $9·8 billion in 1970.

Oil production: due to the combined effect of the world slump and some demand elasticity effects, non-communist oil production actually fell between 1973 and 1975, the second time a fall had been registered since the First World War: 48·2 million b/d 1973, 47·5 million b/d 1974, 42·1 million b/d 1975, 44·8 million b/d 1976. (Chase Manhattan, *Annual Financial Analysis of a Group of Petroleum Companies – 1970, 1975. Energy Report from Chase,* March 1977)

Demand growth predictions: the predictions made prior to the Yom Kippur War of OECD oil demand in 1985 were 75 million b/d; estimates made in early 1977 have reduced the forecast to 52 million b/d – a decrease of some 30 per cent. (OECD, 1974, vol. 1, p. 9; *OECD Observer*, March 1977, p. 4)

What do all these statistics imply? Company apologists make at least two comments. First, they argue that part of the rise in company profits over 1973–4 reflected the fact that companies made some once-and-for-all inventory gains. Some, like the US ones, had substantial non-OPEC sources of crude oil and all had something like three months' supply of crude oil flowing through their systems at the time. The price increases decreed by OPEC in late 1973 and early 1974 meant that the value of these volumes of oil rose, and was shown as an additional source of profit on the balance sheets of the bulk of the majors which had not then switched to 'last-in, first-out' accounting principles which down-play this effect. Texaco claimed that some 24 per cent of their 1973 and 1974 profits were explained by this phenomenon which of course, does not guarantee that the company will realise such profits in the future. Secondly, they argue that company profits have risen nowhere near as fast as government taxation. In the words of a Chase Manhattan writer commenting on the provisional 1976 financial results of the Chase group:

> It is . . . apparent that direct taxes have grown at a faster rate than the other financial categories. When comparing the growth in taxes to the growth of net income, it is indeed obvious that government has benefitted substantially more than the companies. A growth rate in taxes of that magnitude might even be considered by some people to be 'obscene'. (*Energy Report from Chase*, March 1977, p. 2)

Company opponents would argue that even if company profits in 1976 were still some 20 per cent below those of 1974, the companies do seem to have raised the general level of their profits to about double what they were before the 1972–4 profits' explosion. In addition, the critics argue, the companies' return on investment has been in the 13–14 per cent region for three years in a row, in comparison with the years 1965–72 when it obstinately stuck in the 10–12 per cent region.

Arguments about profits are notoriously complex, and judging motives by what happens to profits some three or four years later is unrealistic. In this case, the general figures on profitability disguise a major shift in the geographical balance of company profits. From a lowpoint in 1973 when the USA contributed about 35 per cent of the Chase group's profits, the US contribution reached 58 per cent in 1976, a year in which these companies failed to increase their profits on operations outside the USA. The result is that their return on investment within the USA reached 15·7 per cent in 1976, a figure well above the rates for 1965–73. On the other hand, their return on investment outside the USA fell to slightly under 12 per cent in 1976, bringing the rate back down to the range of levels in the late 1960s. It is possible to argue that the companies knew all along that letting OPEC raise prices would ultimately benefit them by raising the profitability of their North American operations but this assumes the companies had an omniscience which is hardly credible in the absence of evidence that company managers or spokesmen were talking this way in the early 1970s. And I am simply not aware of such evidence. In particular, I can certainly find no evidence to support Peter Odell's arguments that sometime around 1968 the majors took a conscious decision to bring world supply back into a more profitable relationship with demand, nor can I find evidence of any influential executive from the majors indicating that he wanted OPEC to raise prices. (This is not to deny that, by around 1973, there was a sense of resignation that OPEC would succeed in further raising prices. There was considerable discussion prior to the embargo about the impact this price rise might have on the world's financial structure by the end of the decade.)

Of course, any such evidence would be hidden, but if it existed, traces at least should have been uncovered by the Church Sub-Committee when it examined the claim that the majors had explicitly or implicitly helped OPEC to gain control, or by authors like Oppenheim who have documented how the Nixon administration may have encouraged OPEC to raise prices during the 1970s (1976). Many oil industry 'watchers' have an interest in proving the kind of case that Odell would like to see made and, at the moment, it is the lack of evidence to support this case which seems significant.

Any company which encouraged OPEC to raise prices would have been taking a major gamble. It is by no means clear that diversification was sufficiently developed by companies at that time to make it logical to push for higher world energy prices, particularly since the cost of producing new energy sources was obviously going to be very much higher than the 10 cents per barrel for producing Middle Eastern oil, which still only cost the companies an average of $1·05 per barrel in 1970 after all payments to host governments had been made. It would have made sense for the majors to have encouraged high-priced oil if these alternative supplies of energy could support them. However, this was not the picture in the early 1970s. In the case, for instance, of the admittedly crude-rich Gulf Oil, the calorific value of the coal it produced in 1973 came to only 8 per cent of the value of its oil production.[4] It would therefore have been suicidal for Gulf to encourage a policy which priced its oil out of world markets on the assumption that the stimulus thus given to its coal activities could somehow compensate for the loss of income. Even if they had backed such a strategy, they would have found themselves in the precarious position of expanding activities which could be made profitless whenever OPEC decided to reduce prices toward the level of marginal operating costs. Unless companies like Gulf had guarantees in advance that parent governments would underpin high-cost investments against such a threat (the 'floor price' concept), they would have found themselves in an impossible situation.

Another objection to the theory of an alliance between the companies and the hosts is that high oil prices were bound to present the companies with a whole series of unpleasant problems, not the least of which would be unpopularity. The US companies had long been prime targets for the US antitrust authorities, and the combination of large oil price increases and increased diversification into coal and nuclear power was bound to arouse their wrath. Even if the Department of Justice did not strike down such diversification, the companies were moving away from an industry which they dominated for historic and economic reasons into oncs where there were competitors. In nuclear power they would be faced with Westinghouse and General Electric which had already won the lion's share of existing markets. In coal, they had snapped up many of the most attractive coal companies but still faced potential competition from mining or steel companies like Kennecott, Amax and Pittsburgh Steel (Ridgeway, 1973, p. 167). In petrochemicals, the competition would be from traditional chemical companies like Du Pont, Dow and Union Carbide.

The differences in company strategies after 1970 were notably less than their similarities. In general, they all conceded reluctantly in the face of demands from increasingly self-confident host governments and market conditions which did not favour last-ditch measures. They fought such demands by forming a common front against OPEC and trying to weaken OPEC unity. But when the crunch came, they proved to be unwilling to jeopardise their future stake in the oil industries of most key producing countries. They gave ground consistently from 1970 to late 1974 when market conditions allowed them to boycott relatively expensive crude oils. Since then they have been winning price discounts for themselves, have limited their commitments to dispose of unlimited quantities of crude oil in all circumstances, and have started to reduce their liabilities to contribute capital toward host states' planned investments.

But despite signs of new stable relationships, there is still a certain ambiguity in the attitudes of companies towards the host oil producers. Certainly lower oil prices would give a welcome fillip to oil demand, but there is no evidence that any company is going to campaign too strongly to bring this about if offending one of the larger oil-producing host governments could jeopardise their future access to crude oil and, perhaps, bring them problems with inventory losses. There is nothing they can do to stop Kuwait, Venezuela, Saudi Arabia or Norway from restricting production. Moreover, they cannot dictate the policies of importing countries which will determine what the global demand for oil will be in coming years. At the time of writing, oil demand is growing again and we are waiting to see whether the United States, the most crucial importer of them all, is politically capable under President Carter of taking the hard decisions which would lead to significant conservation in its oil use. All the signs, then, are that the market will tighten in the coming decade, and the lessons of 1970/1 and 1973/4 are that in such circumstances the companies lose ground and find themselves in relatively poor bargaining situations. As long as the Saudis are unwilling to break OPEC, it would be a rash executive who advised his company to try leading a boycott of Iranian crude oil in an attempt to drive prices in this one key producing state to levels sufficiently below those of the rest of OPEC members for that organisation to crack. In a situation of glut, some such strategy could just work. In a tight market, it would be suicidal.

As long as the oil business makes up the bulk of their activities, the oil companies are partially dependent on the host producer governments. Given current predictions of increased reliance on Middle Eastern production, a major which ran foul of Saudi Arabia, Iran and perhaps one other host producing state to the extent that these countries refused to sell it crude oil either at or below the market price would be driven out of business. And this is a real risk. Saudi Arabia, in particular, needs no one individual company, but any of the Aramco partners would be grievously affected if it were driven out of this traditional preserve. The power to refuse to sell is a powerful weapon in a perennial seller's market. And no corporate strategist can guarantee that it will become a buyer's market – at least not until we see the extent to which demand for oil is really going to run behind the growth rates of the world's GNP, and the speed with which new oil supplies from areas like Alaska, the North Sea and Mexico are really going to eat into OPEC's traditional markets.

However, corporate dependency on the host producing governments involves more than future guaranteed supplies of crude oil. The petro-dollar reserves of the OPEC economies loom large in the eyes of corporate planners whose job is to predict the buoyant markets of the 1980s. Thus Aramco is managing Saudi Arabia's $15 billion project aimed at utilising the associated gas upon which diversification into industries like petrochemicals and iron ore processing depends. Mobil is involved in the development of the Red Sea port, Yanbu, as a terminal for oil from the eastern part of Saudi Arabia. Mobil is planning an 800-mile pipeline to supply Yanbu and is also negotiating with the Saudi authorities over a fifty-fifty joint refinery and petrochemical complex there. Other oil companies which are potentially involved in Saudi petrochemicals are Exxon and Shell, despite some Western warnings about an excess of petrochemical capacity in the 1980s. Elsewhere in the Middle East, Gulf and the Kuwaiti government are establishing a jointly owned real estate company which will do business throughout the Arab world; BP has joined the National Iranian Oil Company in a fifty-fifty joint tanker venture and Shell is in a thirty-seventy joint venture with the government of Qatar to develop that country's gas potential. In Venezuela, BP has a minority stake in a venture to build a 100,000-ton protein-from-oil plant (*Petroleum Economist*, 1975, pp. 190, 254, 338; *Petroleum Intelligence Weekly*, 3 May 1976, p. 3, 5 April 1976, p. 12, 29 March 1976, pp. 10–12, 15 March 1976, p. 10). By no means will all of such ventures come to fruition, but the implications are clear. The OPEC countries now have sufficient income from their oil exports to justify investment in extensive capital-intensive projects and there will be many areas such as petrochemicals, general development and the management of large-scale projects where the majors will be able to use their advantage of technologies and historic ties. This should mean that they will win more than their fair share of such projects, so they would be mad to antagonise host governments unnecessarily. If the activities surrounding crude oil are currently the backbone of their business, it is by taking advantage of favourable investment opportunities of the size offered by the OPEC countries which will ensure that they do in fact survive into the 1980s and beyond. This is not to say, however, that the dependency is so great the majors will tolerate any behaviour or any situation. Moreover, whether they diversify or stay wedded to oil-related technologies, competition will grow. Smaller oil companies, state-owned companies and companies moving out of hitherto non-competing industries will all serve to limit the majors' freedom. Too-high tax payments will lead to oil being priced out of its markets, and thus to slower-than-predicted growth for oil-based companies. Too-generous terms in helping the host countries industrialise could lead to bankruptcy. Dependency does exist, of course, in the sense that the current and future fortunes of the majors can be harmed or facilitated by OPEC decisions. Company attitudes toward the price of oil are influenced not only by the current state of the market, but also by their hopes for future deals within the oil-producing world. Investing in an uneconomic petrochemical plant, therefore, or diverting scarce personnel into helping industrialisation may be bad business in a strictly book-keeping sense, but justified as gestures which help a company secure long-term supplies of crude oil. And a refusal to fight too hard for lower oil prices may be bad business today, but may help the majors maintain a preeminent position in the economies of OPEC members in the future.

Either way, it is clear that there now is a community of interests between companies and host producer governments which did not exist before the 1970s, but this should not be overstressed. Ultimate power is now with Saudi Arabia which alone has the ability to produce at a rate which could destroy OPEC, but all the evidence since the December 1976 OPEC meeting at Doha is that the Saudis have no intention of following such a course. The companies are impotent bystanders, gaining by a high-price strategy in some areas of their activities (North American oil production, alternative energy developments and helping host industrialisation) but losing in others (participation means that their hold on OPEC crude oil supplies will continue to decrease, and high oil prices mean that energy conservation in the OECD world will shrink their potential markets). The statistics show that the oil companies have on balance gained on most counts from OPEC's pricing policies since 1973, but this can give them little satisfaction. Their vulnerability to government decisions at both the producer and the consumer end has been increased tremendously. Profits which have been given them by government decree can just as easily be taken away.

NOTES

1 Blair gives impressive statistical evidence of the way the majors were once able to orchestrate production within the OPEC world (1976, pp. 99–101).
2 This was part of a series of Mobil advertisements.
3 Chase Manhattan has studied a group of oil companies for about forty years. Currently their group consists of twenty-nine companies which include all the majors.
4 My calculations are from a table in *Petroleum Economist*, October 1975, p. 379.

8
Companies, Industrialised Consumers and the OPEC Challenge

The governments of parent and other industrialised countries entered the 1970s almost totally unprepared for the mounting challenge which was to come from OPEC. They had relied on the majors to provide them with cheap oil and, during the relatively quiescent 1960s, this strategy had been fully justified. The companies were slow to sense that the world environment had changed, and it took even longer for importing countries to join them in a realisation of a major power-shift within the industry. It took the Arab embargo of 1973 to finally stir them from their lethargy. The role of the oil companies was to try to absorb the first shock of the OPEC challenge (not too successfully), to give some warnings to the industrialised world (with little effect), to blunt the worst of the Arab embargo (one of their successes) and, with the blessing of the importing countries, to search for alternatives to oil.

INDUSTRIALISED WORLD'S UNPREPAREDNESS

Until 1969 at least, few forecasters within the majors saw trouble ahead as far as the balance of supply and demand was concerned.[1] The 1960s had caused problems for the industry, but these had primarily been connected with potential overcapacity. There appeared to be no obvious physical limits on increased production from the Middle East, and pressures stemming from the rivalry between Iran and Saudi Arabia were interpreted as ensuring a likelihood of overproduction rather than underproduction in the region. Rapidly expanding production from Nigeria and Libya, where the independents were expanding production as fast as they could, meant that the majors saw their problem as one of discreetly limiting increases in productive capacity at a time when host governments were becoming increasingly aware of the companies' pro-rationing power. A December 1968 memorandum produced within Socal's economics department represents conventional oil-company wisdom of that time:

> In summary, the over-hang of surplus crude avails is very large. Pressures will exist to continue to produce in many areas in excess of market requirements. Our forecasts of production from the Middle East could be high . . . if crude from Libya and Nigeria enters the markets at the higher levels that seem to be not only physically possible but also probable. While the governments of Iran and Saudi Arabia will likely exert considerable pressure to raise production by the Consortium and Aramco considerably above the levels forecast above, we believe it will be extremely difficult to achieve even these volumes and modest growth. (Church Report, 1974, pt 7, p. 363)

Socal subsequently stressed that this document was a voluntary think-piece and its conclusions were speculative. However, it appears to be much in line with industry thinking around this time (Church Report, 1974, pt 7, p. 371). Exxon's forecasts concerning Middle Eastern oil made similar assumptions about the difficulties of expanding production at the rates experienced in the 1960s. The growth in demand for crude oil in the Eastern hemisphere was decreasing and the increase in non-Middle Eastern oil production was eating into the market for Middle Eastern crude oil.

> The result is an inordinately low growth available for the Middle East as a whole that must be further divided among the new Middle East concessions as well as the established producers. No known method of allocating the available growth is likely to simultaneously satisfy each of the four major established concessions, i.e. Iraq, Iran, Kuwait and Saudi Arabia. (Church Report, 1975, pt 8, p. 606)

Sadly for the importing countries (and the oil industry's forecasters) practically everything that could go wrong with these predictions did so. First, they underestimated worldwide demand for oil. Economic growth was faster than they had assumed, especially in Japan; environmental protective measures added to energy consumption (smog controls, for instance, made cars more petrol-thirsty), and a faster-than-predicted decline in coal production put further pressure on oil demand (US environmental restrictions on sulphur emissions, a fall-off in European subsidies to coal, and higher costs stemming from the 1970 US Mine, Health and Safety Act all contributed to the coal shortfall). By 1973, the non-communist world was using some 4·7 million b/d oil more than Socal's 1968 forecast which also proved too optimistic concerning the production levels in Libya and elsewhere. Libya ordered production cutbacks in mid-1970 for negotiating reasons (there was also genuine concern about overproduction from its reserves) and by 1973 Libyan output was 2·2 million b/d below expectations. Kuwait and Venezuela also adopted conservation measures with similar, though less spectacular results. Predictions for US production went noticeably wrong as well, with a 850,000 b/d shortfall in 1973 below the predicted level. Both Exxon and Socal have explained this shortfall by claiming that they were too optimistic about the potential of offshore production, adversely affected by the 1969 Santa Barbara disaster which delayed the federal leasing of further offshore areas. The Alaskan pipeline was also held up for environmental reasons. (Church Report, 1974, pt 7, pp. 331–4, 347–50, 360–3, 370–3.)

For a variety of reasons, the Middle East came to play a completely different role. Instead of fighting a losing battle to maintain its hold over Eastern hemisphere markets, it became even more central to them. The combination of buoyant world demand and production shortfalls meant that even doubling Saudi production levels still left an uncomfortable shortage of productive capacity. From a peak of spare crude oil capacity in 1965 of around 7 million b/d the world was faced with less than 1 million b/d in 1973, in line with the much increased overall demand. Importers were vulnerable to any kind of supply shutdown – and the industrial world was singularly ill-prepared for shortages (Church Report, 1974, pt 7, p. 335). This was not very surprising since the importing governments generally did not have any units charged with (or capable of) double-checking industry forecasts. This meant that the outlook of government officials reflected that of the oil majors over the state of the oil market. The relevant officials mostly assumed that overcapacity was the problem; those who thought otherwise were few and not influential enough to sway government policies.

Conventional thinking saw the chief threat to the stability of the industry as coming from the Arab-Israeli conflict which had twice interrupted oil-flows. In the aftermath of the 1967 Suez War, the industrialised world had taken the problem seriously, and some contingency planning was done. However, policy makers gradually grew complacent, particularly as the price increases conceded during the Teheran-Tripoli negotiations came to be seen as less unsettling than had been feared. The need to plan for a new embargo or a major OPEC drive for drastically high price levels seemed to recede. By 1973, when the oil industry belatedly came to grasp just how critically tight the market was, and how seriously key Arab producing states viewed the Israeli issue, it found itself trying to warn politicians wrapped up in their own complacency. The oil embargo came too quickly for the industry to get across the message that it was not just the political relationships within the oil industry which had changed, but its underlying economic realities.

Though some governments did have some form of oil or energy policies in the early 1970s, these were generally ill-thought-out affairs, reflecting the relative strength of domestic pressure groups, rather than rational responses to an impending energy crisis. In the USA, there was a review of the import quota policy (see p. 51) in 1969–70 after the election of Nixon who seems to have been motivated less by awareness of impending problems than by a suspicion of the oil companies which had been close to his Texan predecessor, Lyndon Johnson. The report from his Cabinet Task Force (the 'Schultz Report') was presented in January 1970, and came to conclusions even more removed from reality than those prepared by the oil companies. The quota system, it reported, could be safely dismantled and replaced by a tariff; the price of domestic US oil could be cut by 80 cents a barrel over three years; none of this would affect domestic exploration, and US reliance on imports from the Middle East could be kept to an acceptable limit of 10 per cent (Rifai, 1972, pp. 1227–9; McKie, 1975, pp. 78–90). The Task Force membership had been deliberately chosen not to rely on oil company experts, and it seems to have relied too heavily on an optimistic report from the Department of the Interior which misjudged domestic production potential. The industry naturally attacked its conclusions which were never implemented, though the import quota scheme was gradually relaxed when it became clear that the US economy needed a higher level of oil imports than hitherto necessary. In retrospect, the timing and optimism of this report was supremely ironic, since 1970 was the year in which domestic US oil production peaked, and in which escalating Libyan demands were sounding the knell of the era in which the majors could control the international industry for the benefit of the Western economies.

Few in Washington felt that the Arab hold on Middle Eastern oil posed much of a threat to the USA. There was concern lest another Arab-Israeli war might trigger an effective oil embargo against the more vulnerable European nations and Japan, but it was felt that this danger could be neutralised by stockpiling and emergency allocation policies co-ordinated by the OECD. There was also a belief that arrangements with Iran could be used to thwart the excesses of its Arab neighbours.

The Europeans had become conscious of their reliance on imported hydrocarbons and, during the 1960s, developed a common approach to stockpiling, an obvious precautionary measure. Some countries had such policies already in existence; French policy originated in the

1920s, while the British had an unofficial understanding with the oil companies that they should stockpile after the Suez crises. The first cautious steps in joint stockpiling were taken in 1958 and 1962 by the OEEC/OECD which set the standards adopted by the EEC in its 1964 Protocol Agreement on Energy, according to which Community governments took steps to build up minimum stocks based on the previous year's transaction levels (sixty-five days' worth in the case of crude oil imports). A 1968 directive required all member states to keep stocks of at least sixty-five days, while in 1971 this was extended to ninety days to come into force in January 1975 (Shell, 1972, pp. 3–4; Mendershausen, 1976, p. 25). In addition, the Europeans also supported an international allocation scheme similar to that which had successfully minimised the disruption caused by the 1956 Suez crises when the rescheduling of oil flows had been co-ordinated through the OEEC, working closely with the industry to share shortfalls between member countries. The majors' key role in this process was recognised by the creation of the temporary body, OPEG (OEEC Petroleum Industry Emergency Group), which brought together Shell, BP, CFP and the US majors as a council to administer OEEC guidelines. In the words of the OEEC study, 'The close association of sovereign governments and an international industry . . . in an operational capacity was breaking new ground' (OEEC, 1958, p. 34).

A precedent for OPEG could be found in the Petroleum Board of the Second World War (see p. 39). Once the formula had been tested in 1956, the spirit of OPEG remained, even if US antitrust laws meant that any such body involving US companies could only be a temporary affair, to be disbanded as soon as its specific task was fulfilled. In 1967 the US authorities gave permission for American companies to join an OPEG successor, the International Industry Advisory Body (IIAB), but French wariness meant that the role of the IIAB in this crisis was restricted to specifically technical matters.[2] Nevertheless, the OECD decided it was worth keeping such an organisation in reserve for some future supply crisis although the emergency scheme could only be activated with unanimous agreement (abstentions were allowed).

Such contingency planning did not consider Japanese and American markets, even though Venezuelan oil, which had been of particular importance to Europe in 1956, was increasingly needed for North American markets in view of the insufficiency of US domestic production. US representatives had stressed this new situation within the confines of the OECD Oil Committee in late 1969. But it was generally felt that there was potential growth left in domestic US crude oil production and, furthermore, that the Canadians could be trusted to provide the necessary extra flows of oil and gas. And even in early 1972, the USA was still fending off the Venezuelans who wanted to be granted preference over Eastern hemispheric suppliers. The US reply that it chose to base import policy on Canadian rather than Venezuelan oil – a rebuff to a traditional supplier – was hardly the action of an administration aware of the magnitude of the challenge which it would face in the next year (Tugwell, 1975, p. 135).

RESPONSE TO THE TEHERAN-TRIPOLI AGREEMENTS

The USA did not initially see Qadafi's overthrow of the Idris regime as a threat to the traditional structure of the oil industry and acquiesced in the closure of the Wheelus airbase in the hope that this would satisfy Libyan nationalism and reduce the threat to American investments (Church Report, 1975, p. 121). Libyan oil diplomacy was primarily a case for the State Department and, in particular, its Office of Fuels and Energy, then headed by James Akins. A forceful personality, Akins's reactions to the uncompromising initial response of the oil companies to Libyan demands in 1970 had some of the ambivalence of State Department reactions to the disputes four years earlier between the Consortium and Iran. In particular, he felt that the Libyans had a good case for a higher price for their oil. He had a very clear idea of the extensive corruption of the previous regime and, like others in the State Department at the time, blamed at least some of the oil companies for having significantly contributed to this. He also believed that the quality of Libyan crude oil and its proximity to European markets justified some form of price premium. He felt the initial Libyan demand of 40 cents a barrel was 'quite reasonable', if not too low, and advised the industry that the Libyan demands were unexceptionable, explaining his motives thus after the event:

> it was, I thought, to our interest overall to get the oil for a fairly low price. It was also to our interest, I thought, that the companies have a reasonable working relationship with the Libyans and with the other producers. If the Libyans concluded they were being cheated, this, I thought, guaranteed a breakdown in relations with the companies and all sorts of subsequent problems. (Church Report, 1974, pt 5, p. 61)

The companies chose not to listen to him, and Exxon took a hard negotiating position which was undermined when Occidental caved in to Libyan demands in September 1970 (30 cents a barrel on the posted price, a five-year deal, annual increments and an increase in the tax rate from 50 per cent to 58 per cent of the posted price). At this point, there occurred one of those occasions which in retrospect look as though firm action by the rest of the industry might have checked host government militancy in its initial faltering stages. Could the majors have held out against the Libyans and, more important, could they have staved off parallel demands from elsewhere?

The industry turned toward its parent governments for support. The Secretary of State, William Rogers, met the US companies three days after Occidental's surrender. A fortnight later there was a further meeting with Under-Secretary of State Alexis Johnson, at which BP and Shell were in attendance as well (they had been canvassing the British Foreign Minister, Sir Alec Douglas-Home). The issues were apparently simple. Shell, BP and Mobil argued that a stand had to be made, and that the industry could do without Libyan oil and still supply Europe with 85–90 per cent of its needs for at least six months. The other companies were less convinced – the European leaders whom Sir Alec approached were not enthusiastic about the possibility of a cut in oil-flows and the general State Department line seems to have been that a Middle Eastern settlement would take care of the oil problem. The specific argument of Akins (see p. 141) was that whatever happened in Libya, Saudi Arabia would not try to imitate (Sampson, 1975, pp. 213–15). In practice, each company went its own way; Socal and Texaco accepted terms similar to those that Occidental had conceded and the other companies had to follow.

The first skirmish lost, the industry was immediately faced with a flurry of demands and the year ended with OPEC creating a Gulf Committee, consisting of oil producers around the Gulf, which was instructed to start negotiations with the relevant companies within thirty-one days, and report back to OPEC within seven days of an agreement so that OPEC could pass judgement on it. The companies were thus faced with the probability that any deal they struck in the Gulf would lead to leap-frogging demands from elsewhere; Libya put this process in motion in early January 1971 with an insistance that terms won the previous autumn be improved.

The companies concluded that they should aim for a joint negotiating stance, and for a 'safety net' of alternative supplies. They found the US government happy to help in so far as both schemes depended on getting antitrust clearance from the Department of Justice in the somewhat ambivalent form of Business Review letters. By granting these, the US administration permitted the creation of the London Policy Group and the Libyan Producers' Agreement.

London Policy Group

The LPG (London Policy Group) was established on 20 January 1971 to co-ordinate the negotiations with the Gulf States in Teheran and with Libya, as representative of the Mediterranean countries, in Tripoli. It was anticipated that these negotiations were going to be inter-related, lengthy and extremely complex, and therefore there was a need for a steering group of senior executives from all the participating companies which could react quickly to new developments, provide answers to queries from the negotiating teams in the field, and modify the terms of reference for the latter, as and when the circumstances seemed to dictate a change in strategy.

The LPG was based in London, eventually in BP's headquarters, but as the top US executives needed to stay in America for most of the time, a corresponding high-level group was set up in New York, eventually in the Jersey headquarters, to provide back-up technical advice and policy decisions. The chief executives at the very highest level were to meet eleven times before the Tripoli Agreement was finally hammered out. In addition there were sub-committees to analyse the question of transport differentials which were of particular importance to both sets of negotiations, to carry out more general economic evaluations of the various proposals and counter-proposals and to handle public relations and, in New York only, supply issues should the Tripoli talks break down. There were also several *ad hoc* meetings to consider tax and legal angles related to the agreements.

The LPG was very much an oil company affair. It brought together the majors, including CFP, all the relevant US independents and a number of non-Anglo-Saxon companies such as Gelsenberg from Germany, Hispanoil from Spain, Petrofina from Belgium. The Japanese AOC (Arabian Oil Company) associated itself with industry's joint message to OPEC which preceded the formation of the LPG, attended some of the Group's meetings, but did not actually sign the Teheran or Tripoli agreements. It did substantially accept their terms when it made its own deal with Saudi Arabia and Kuwait in May 1971. The only two notable stand-outs were the state companies, ERAP and ENI, which argued that their interests were distinct from those of the majors – though this did not actually win them favourable treatment in the months and years which followed (Church Report, 1974, pt 6, pp. 234–42; Tugendhat and Hamilton, 1975, p. 118). As the two sets of negotiations intensified (the agreements were signed in Teheran on 14 February and in Tripoli on 20 March 1971) the LPG met almost daily, with its New York counterpart meeting only slightly less often. No government officials took part directly, though Akins based himself in London until the end of the Teheran negotiations and daily met involved company representatives, and the British government was well placed to keep in touch. The nearest there was to a government presence was the regular attendance of a partner or associate from the firm of Milbank, Tweed, Hadley & McCloy, the US companies' legal adviser whose job it was to ensure that the discussions remained within the areas indicated to the US Justice Department (Church Report, 1974, pt 5, p. 18, pt 6, pp. 236–71).

The original intention was to dismantle the LPG machinery as soon as the Teheran-Tripoli negotiations were over, and indeed, its members voted to do this in July 1971, leaving in its place an Administrative Group which was to limit itself to questions springing from the interpretation and implementation of the two five-year agreements. However, hardly had John J. McCloy, the companies' redoubtable legal adviser, informed the US antitrust authorities of this decision, than he was back before them asking for another Business Review letter to cover the LPG's reactivation, in response to OPEC's demands for participation in existing concessions and for upward adjustment of oil revenues to offset the effects of the currency turmoil following the suspension of the dollar's convertibility to gold in August 1971. The Department of Justice obliged and the companies were permitted to take a joint stand during the participation negotiations and the Geneva and Supplemental Agreements of January 1972 and June 1973 which covered currency readjustments. In September 1973 they were back once again for Justice Department permission for a communal industry approach to OPEC's demand that the Teheran and Tripoli deals be substantially re-negotiated. Once again, a Business Review letter was forthcoming, but the negotiations in Vienna were overtaken by the drama of the Arab oil embargo. At that point, the kind of inter-company co-operation carried out in the LPG became more crucial than ever – though the documentation of how it worked after the embargo is sparse (Church Report, 1974, pt 6, pp. 223, 270).

Libyan Producers' Agreement

Libya's ability to pick off the smaller companies was the Achilles' heel of the oil industry's attempts at a joint negotiating strategy. The companies had therefore to find a way to help any one of their number which might be singled out by the Libyan authorities for enforced production cutbacks or even outright expropriation.[3]

The industry's answer was the Libyan Producers' Agreement, or the 'safety-net agreement', which specified that the signatories would replace any shortfall suffered by a company as a result of resisting Libyan demands for the good of the industry. The alternative supplies would come from other production in Libya and from crude oil produced in the Gulf. It would be supplied in proportion to a company's share in Libyan production, with transferred crude oil made available at, or near, cost. This agreement was drawn up at the same time as the LPG was formed and was accepted by the US Justice Department on 15 January 1971. There were fewer companies involved than in the LPG, reflecting the fact that not all companies with international oil interests were producing crude oil in Libya. American and Anglo-Dutch companies predominated, though Gelsenberg was actively involved in the scheme from the beginning, and Hispanoil and Japan's AOC both subscribed at various stages (Church Report, 1974, pt 6, pp. 223, 271–88).

Production cutbacks were not the only threat facing companies producing crude oil in Libya, as BP found in December 1971 when its

Siris Field concession was nationalised, triggering the crude oil sharing envisaged in the arrangement. Bunker Hunt and Occidental were forced to take major production cutbacks in the following year, whilst most of the other Libyan producers had to accept smaller ones. By December 1972, the shortfall was about 650,000 b/d and BP, Bunker Hunt and Occidental were the major recipients of diverted crude oil. The time-span of the agreement, which had been strengthened to cover the effects of nationalisations, was further extended in November 1972 in response to Libyan pressures for participation following a participation deal with AGIP. The Libyan government nationalised Bunker Hunt in June 1973 and took a 51 per cent stake in all the other companies operating in Libya by September. All told, some 240 million barrels of oil were reallocated under this agreement up to the end of April 1974, with Occidental receiving 77 million barrels, Bunker Hunt 51 million and BP (also a major contributor of Gulf crude oil to be shared) taking 59 million (Church Report, 1974, pt 6, p. 272).

The scheme ran into trouble in 1973 when the oil market tightened, and crude oil for sharing could only be produced at the expense of lost sales for the supplying company. Despite having no Libyan interests, Gulf Oil had joined the agreement because it accepted the need for industry co-operation in the face of host government threats. But when Gulf's major source of non-American oil in Kuwait was cut back by government decree, Gulf chose to discharge its obligations by the option of a payment of 10 cents for each barrel of crude oil owed. After a dispute over the calculation of its obligations, Gulf ended up supplying a mere 3·4 million barrels, a quarter of the contribution of Mobil, the next least-involved major (Church Report, 1974, pt 6, pp. 269–70, 272).

In a sense the agreement failed, since Occidental, Amerada Hess, Conoco and Marathon all signed agreements in August 1973 accepting 51 per cent government participation in their operations. On the other hand, it fortified the resolve of Bunker Hunt which refused to give in, insisting that it would only negotiate on the basis of participation terms agreed to by the companies in the Gulf. The result was that Bunker Hunt was fully nationalised in June 1973. It is not clear if the independents signed their eventual deals with the Libyans with the reluctant blessing of the other participants in the Libyan Producers' Agreement (they probably did), but it was the kind of defiance shown by Bunker Hunt in 1973 which had been so lacking in 1970. They certainly could have come out worse in 1973. The Iraqis had already gone for 100 per cent nationalisation, so it was surprising that the relatively fearless Libyans should have settled, in general, for a 51 per cent stake. Moreover, Libya subsequently ceased to be in the vanguard of the OPEC movement for greater participation.

What happened in Saudi Arabia and Kuwait became far more important, and the fact that Libya settled for a less abrasive role may well have been the result, at least partially, of the united front the industry presented under the Libyan Producers' Agreement.

Uniqueness

As industry institutions, the LPG and the Libyan Producers' Agreement were unique in the sense that the companies had never before grouped together to negotiate with governments. Antecedent bodies had either been formally created to help parent governments handle the logistic problems of the Second World War, the Korean War and the 1956 and 1967 Suez Wars or else had been private arrangements like the prewar 'London Committee' which had discreetly co-ordinated the activities of the companies subscribing to the Achnacarry Agreement and its succeeding arrangements from a British base safely outside the reach of the US antitrust authorities. The other combinations of oil companies were the oil-producing consortia which were merely concerned with organising oil production in a single country or a narrow range of countries (Church Report, 1974, *IPC*, pp. 8, 40). The LPG was the first governmentally approved, inter-company bargaining body. In addition, both the LPG and the Libyan Producers' Agreement were distinctive in the breadth of the companies they involved as antecedent groupings had primarily concerned only the majors. The oil world had changed. Within the Anglo-Saxon world, the US independents which had been granted a restricted entry into Iran in the aftermath of the 1951–4 crisis were now given a responsible role in industry councils as a reflection of their growing importance (though it was their overdependence on Libyan production which had caused so much trouble). Amerada Hess, Ashland, Arco, Conoco, Grace, Bunker Hunt, Marathon, Murphy and Occidental were brought into one or both of these new key groupings[4] and Henry Schuler of Bunker Hunt was one of the most important witnesses in the subsequent hearings of the US Senate Committee on Foreign Relations on the oil companies and foreign policy (Church Report, 1974, pt 5, pp. 75–101, pt 6, pp. 1–60), giving his critical views of how the negotiations surrounding the Teheran-Tripoli Agreements had been handled.

The LPG and the Libyan Producers' Agreement signified the end of the era in which a handful of Anglo-American, Dutch and French companies were all that mattered to the industrialised world. Gelsenberg, for instance, played a prominent role throughout this period, providing one of the six main negotiators in Tripoli. It was one of the thirteen companies which decided that the LPG needed reactivating in the light of OPEC's October 1971 demands regarding participation and currency readjustments (CFP was the only other company from outside the central Anglo-American-Dutch core). The fact that AOC from Japan was also a member meant that there were now representatives from the two most important non-parent economies (reflecting, in the case of Germany, a return to a role in industry councils which it had been on the verge of assuming at the outbreak of the 1914–18 War). These two bodies represented a new stage in the evolving relationship between oil companies and their parent governments. For the first time the latter realised that the companies could no longer resist host government demands, and that traditional support would no longer be sufficient. Nevertheless, the parent governments still held back from entering direct negotiations, preferring to bolster the companies' bargaining strength through bodies like the LPG. They chose to restrict their role to monitoring the industry's activities at second hand, with officials like Jim Akins making only limited inputs into industry councils. The original antitrust clearances given in 1971 show clearly that the US authorities saw little need at the time for permanent bodies of this type. The LPG and the Producers' Agreement were seen as institutions which would be disbanded as soon as the immediate negotiations were completed. It was clear to Akins, however, that the industrialised world was now faced with a continuing crisis and, in December 1971, he floated the idea of creating a permanent advisory committee of US oil companies (majors and independents) to advise the State Department on official US policy. The companies were sceptical (the scheme was too rigid, the companies would be seen as instruments of US foreign policy, etc.) and the idea was shelved (Church Report, 1974, pt 6, p. 303). Given the severity of US antitrust regulations, it was always difficult to persuade executives from different companies to meet jointly with State Department officials, for fear of how the Justice Department might react.[5] Without an advisory committee such as that suggested by Akins, the State Department probably had less effective liaison with US companies than its counterparts in Britain, Netherlands and France had with theirs. It came to rely quite heavily on John McCloy, the veteran diplomat and lawyer[6] who was, by 1973, representing the interests of at least twenty-six of the world's oil companies. It was he, for instance, who first raised with President Kennedy the need for some form of industry concertation against OPEC should that body ever start to flex its muscles (Church Report, 1974, pt 5, pp. 255*ff*.). McCloy warned each new Attorney-General from Bobby Kennedy onwards about the potential need to give antitrust clearance should some such co-operative approach be needed. It was he who guided the industry's plans for the LPG and Libyan Agreement through the legal shoals in 1971. His law firm monitored all their meetings and his routine liaison with the Justice and State Departments provided the main conduit of information between the US government and the oil companies. This does not mean that the State Department could not consult formally with groups of oil industry executives, but calling such a meeting normally involved going through McCloy, and the central institutions at the heart of the industrialised world's strategy (such as it was) against OPEC demands remained overly dependent on the goodwill of the US antitrust authorities. By September 1973 the Department of Justice was starting to grow uneasy about the way apparently temporary bodies like the LPG were showing signs of becoming permanent fixtures. When the industry asked for a Business Review letter to cover its October negotiations with OPEC in Vienna, the antitrust authorities this time insisted on interrogating the individual US companies involved, and the resultant Review letters were even more heavily laced than usual with qualifications and implied warnings that the industry should not take the Justice Department's benevolence for granted.

> We feel it necessary to emphasize, however, that we view such intimate co-operation among high executives of competing oil companies concerning crucial factors involving cost and supply as potentially raising serious antitrust dangers. Accordingly, we view this as an area which we must subject to constant review and reevaluation . . . Our non-disapproval is in no way intended to sanction or authorize any joint oil company action which tends to reduce the supply of petroleum to the United States, such as joint agreements with OPEC concerning production levels or refinery construction, or joint agreements among oil companies to halt production or cease lifting oil in any country, to boycott oil from any country, or to chase so-called 'hot oil'. (Church Report, 1974, p. 265)

However, the Arab oil embargo and subsequent search for increased security of oil imports to the industrialised world made international co-operation amongst the oil companies more essential and required governments to play a more active and formal role than they had intended.

CONSUMER GOVERNMENTS' STRATEGY – HOW EFFECTIVE?

There are those who argue that the strategy whereby the companies were given parent government blessing to negotiate jointly with OPEC was an unmitigated disaster. In particular, blame is laid at the door of the US State Department which allegedly failed to give the companies the strong diplomatic backing they so badly needed in 1970–1. Senator Church's Sub-Committee devoted considerable attention to the role of the State Department in January and February 1971, arguing that the mission of Under-Secretary of State John Irwin to Teheran on 17 January pulled the rug from under the feet of the oil companies, making it impossible for them to hold one single set of negotiations with the whole of OPEC (Church Report, 1975, pp. 128–34).[7] This argument was summarised in a major *Forbes* article in April 1976 entitled 'Don't blame the oil companies: blame the State Department'. Others like Morris Adelman have pointed to the OECD meeting which the USA convened on 20 January, to assure the Europeans that acceptance of OPEC demands would guarantee at least five years of stability. 'This was an advance capitulation', Adelman argued. 'The OPEC nations now had a signal to go full speed ahead because there would be no resistance' (1972, pp. 80–1). Substance is given to the charge by hindsight of these five years of alleged stability. The ink was hardly dry on the Teheran and Tripoli Agreements when the companies were again forced to bargain on participation and currency adjustment issues, and OPEC demands in the autumn of 1973 for a complete re-negotiation of host government tax take came only two years after the signing. Likewise, the Arab oil embargo and the massive price increases of late 1973 and 1974 can hardly be counted as evidence of 'stability'.

The picture might have been different if it had been possible to isolate events in Libya and thus prevent a chain reaction throughout OPEC. This would have required an immediate, concerted strategy from the oil industry as soon as Qadafi's first demands were made, but because of tensions between the majors and the independents and the unexpectedness of the source of the trouble, there was a fatal delay in formulating a joint approach. US officials believed that no industry embargo of Libya would work because it would hit Europeans too hard. Germany would have been particularly badly affected as in 1968 some 43 per cent of its imported oil came from Libya (Mendershausen, 1976, p. 31). It is also possible that the State Department was worried that crude-short European governments would sign bilateral deals with Libya at the expense of the US companies. Jim Akins's own published explanation suggests that both considerations were at work:

> The main reason for not following this course [daring the Libyans to nationalise] was the fact that the loss of all oil from Libya alone would have meant the drawing down of more than half of the European oil reserves within a year. It seemed unlikely, indeed inconceivable, that France, Germany, Spain or Italy would have allowed that to happen; especially as the goal would apparently have been only to protect the Anglo-Saxon oil monopoly, which they had long sought to break. To have tried to explain to them

that they would themselves suffer in the long run, would have been less than futile. We in the State Department had no doubt whatever at that time, and for those particular reasons, that the Europeans would have made their own deals with the Libyans; that they would have paid the higher taxes Libya demanded and that the Anglo-Saxon oil companies' sojourn in Libya would have ended. (1973, p. 471)[8]

In the circumstances it was inevitiable that longer-established oil-producing governments like those of Saudi Arabia, Iran and Venezuela should have wanted to keep pace with Libyan gains. A key personality was the Shah of Iran, who proved himself a rough, tough negotiator with the oil industry from 1966 onwards. Although not as motivated as Qadafi by a concern with Palestinian affairs and rather more aware of wider strategic issues, the Shah had the majors very much on the defensive by 1971 and it was probably only a matter of time before Iran made a similar kind of negotiating breakthrough. The Libyan success merely accelerated the process. Nevertheless, the State Department's diplomacy immediately preceding the Teheran negotiations was somewhat hesitant. In January 1971, when the LPG and the Libyan Producers' Agreement were being formed, the industry asked the State Department to give it some diplomatic backing. The result was the Irwin mission to the allegedly moderate states of Iran, Saudi Arabia and Kuwait in which Irwin sought to put across the US position on the coming negotiations. The oil companies had impressed upon the State Department the extent of their worries about the danger of 'ratcheting' between the Libyans and the Gulf states. Their intentions were expressed in a joint letter to OPEC:

We have concluded that we cannot further negotiate the development of claims by member countries of OPEC on any other basis than one which reaches a settlement simultaneously with all producing governments concerned. It is therefore our proposal that an all-embracing negotiation should be commenced between representatives of ourselves . . . on the one hand, and OPEC as representing all its Member Countries on the other hand, under which an overall and durable settlement could be achieved. (Church Report, 1975, pp. 127–8)

The first country on Irwin's schedule was Iran. During the course of negotiations with the Shah and his finance minister, Amouzegar, it became clear that the Iranians were hostile to the idea of a single set of negotiations, relying partly on arguments (the radical governments would set the tone if OPEC bargained as one) and partly on threats (it is a little unclear, but according to US ambassador Douglas

MacArthur, either the Shah or Amouzegar called the joint negotiating strategy 'a dirty trick' and threatened to stop the flow of oil from the Gulf).[9] In the end Irwin was convinced that the Gulf states were willing to promise not to leap-frog if Libya achieved a better deal. He recommended to his superiors that the industry should negotiate separately with the Gulf states, thus implicitly suggesting that it drop its insistence about negotiating with OPEC as a whole. Secretary of State Rogers accepted Irwin's reading of the situation, and the two industry negotiators who followed Irwin to Teheran found themselves trying to stick with their original terms of reference, while Amouzegar was telling them that their parent governments were no longer committed to the global approach and would be quite happy to let the companies settle for a regional or Gulf approach (Church Report, 1975, pp. 131–2). The company representatives were obliged to accept this view and could only hope that it would be possible to keep the two sets of negotiations 'separate but necessarily connected' (Church Report, 1974, pt 6, p. 22). In practice, the settlement in Teheran was not conditional on a satisfactory settlement in Tripoli – or vice versa. Governments like the French, Belgian and Japanese, which had no stake in Libya, then had no direct interest in what happened in the Tripoli negotiations, according to Bunker Hunt's Henry Schuler (Church Report, 1974, pt 6, p. 23).

Schuler is the strongest critic (though an interested party) of the State Department's *volte face* on a common approach to OPEC and, as a front-line negotiator in Libya, his views deserve some attention. He argues that over the critical period of 16–24 January 1971 his own independent company, and others, managed to live through three consecutive ultimatums from Major Jallud, the Libyan Deputy Prime Minister. The fact that the vulnerable Bunker Hunt was able to survive without even the threat of a shut-in, Schuler argues, should be taken as evidence of the importance of solidarity in the industry as well as of the fact that much of the Libyan bravado at that time was still bluff. He points to the potentially restraining force of Algeria, then keeping clear of any firm connection with OPEC and involved in negotiations on finance and markets with American interests which it did not want to jeopardise by antagonising the US government. Finally, Schuler argues, the Libyans were apparently willing, at one point, to accept OPEC-wide negotiations except for some peculiarly Libyan problems. Had the industry stood firm in Teheran and insisted, with State Department backing, that there would only be one set of negotiations, the Libyans might not have proved intractable (Church Report, 1974, pt 6, p. 11). In Schuler's words:

> Perhaps the greatest consequences of the abandonment of joint negotiations are the least tangible. At the commencement of the

> two week period, there was a certain air of caution on the part of the Algerians and Libyans as well as the Venezuelans and Iranians, on the other hand; there was a certain air of confidence on the part of the companies that, for the first time, all companies were ready and willing to work together towards a reasonable settlement. By January 31st these roles were reversed; the OPEC countries were confident of their ability to face-down the oil companies, home governments of those companies and consumers governments; and the companies had reverted to an attitude of narrow self-interest. This was the start of a rapidly accelerating momentum which has brought us to the point where we find ourselves [mid-1974]. (Church Report, 1974, pt 6, p. 23)

Although Schuler was characterised by McCloy as 'one of the hard-nosed people' involved with these negotiations, and his views have not been publicly accepted by the rest of the industry, his reading of the psychology of the period should not be disregarded. There is evidence of some initial disappointment within the oil industry that the State Department was not going to back their insistence on industrywide negotiations. Three years later McCloy, though disassociating himself from Schuler's outright criticism, commented 'We weren't too much impressed, if I may say so, by the attitude of the US Government' (Church Report, 1974, pt 5, p. 266). And George Piercy, a senior vice-president of Exxon, also criticised the State Department's decision: 'I was disappointed because I thought it was going to make my life harder. I didn't think it was catastrophic . . . because there is always more than one way to solve a problem, but I thought it was going to be harder, much harder' (Church Report, 1974, pt 5, p. 221). On the other side of the bargaining table, Taki Rifaï, petroleum adviser to the Libyan government, claimed: 'It was the starting point of a new era, an unprecedented total victory for the producing countries on all fronts. The oil companies offered no resistance. The companies were there to sign, not to fight' (*Forbes*, 1976, p. 76). It is doubtful, though, that events in succeeding years would have been much different if a tougher and more unified position had been imposed on OPEC in 1971. The devaluation of the dollar in December 1971 had an impact similar to that which resulted from the oil companies' cut in the posted price in 1959 and 1960. Both led to a drop in host governments' income. The events of December 1960 led to the formation of OPEC; the events of December 1971 would have inevitably led to a re-negotiation of whatever financial terms the oil industry was then working under – and this would have occurred regardless of whether or not the companies managed to restrict their concessions to the host governments to a minimum in 1971.

The participation issue is a little less clear. Until the end of 1971, when Libya expropriated BP for political reasons, the only recent example which came close to being a successful expropriation within the oil world was Algeria's decision in February of that year to take a 51 per cent stake in the operations of the French oil companies. Given the fact that Algeria was traditionally a French oil province and that Paris showed no signs of wanting it otherwise, there was little or nothing that the Western powers could do to help. The Algerian decision was bound to be imitated by Libya unless Libya was badly mauled in the negotiations in progress and then a ripple effect through the more militant oil producers like Iraq to more moderate countries like Saudi Arabia would probably be inevitable. Given the central importance of the oil industry to these economies, the desire for participation was bound to increase in the same way that nationalisations and participation were starting to occur in the copper and bauxite industries elsewhere. Defenders of the Teheran-Tripoli Agreements point out that they were concerned only with price and that participation was not something the oil industry had hoped to forestall by these negotiations.

A far tougher line by the industry in 1971 might have made OPEC hesitate before insisting on a premature revision, but then there were more than enough events after 1971 which would necessarily have increased both OPEC's determination and optimism by the autumn of 1973. The combination of the commodities boom of 1972–3, the anti-Americanism of the Arab oil producers in the run-up to the Arab-Israeli War, the growing influence of Third World radicalism, the tight oil market and Libya's success in winning substantial participation were, combined, more than enough to guarantee that the OPEC of late summer 1973 was going to be difficult to control, however well it might have been handled in early 1971. In the light of such circumstances, it would have needed extraordinary diplomacy to withstand OPEC demands for revision. But, in fact, Western governments never succeeded in developing in 1973 even the kind of concerted diplomacy which was vaguely seen in early 1971. The most crucial actor, the US government, retreated and became more remote from the industry, leaving the companies to bear the brunt of OPEC demands at a time when the situation demanded more, not less, parent government involvement.

One factor inhibiting parent governments in their attempts to withstand new OPEC pressures was that there was no guarantee that other governments saw the issues as starkly as they did. To governments like the American, British, Dutch and French, support for their companies was seen a way of defending consumers in general. To non-parent governments like the Japanese, Italian and West German, the situation was more ambiguous and disputes with the OPEC countries often appeared as cases in which parent governments intervened to maintain the commercial pre-eminence of the majors. There seemed little reason why they should get involved in disputes in which their oil supplies might be disrupted just to maintain the position of corporate interests in which they had no stake. Hostility to the majors was nothing new in Europe, and Akins reported one European minister in 1967 as having said: 'American companies brutally conquered our market; if they do not keep us supplied at all times, they will be expelled' (1973, p. 472).

Parent governments kept in touch during the 1970–1 crisis. The links were probably closest between the British and the Americans, but the Dutch and French were kept informed as well. The US ambassador, for example, briefed his British, French and Dutch colleagues on Irwin's trip to Iran, giving the Dutch a chance to object to the apparent drift in the State Department's thinking (Church Report, 1974, pt 6, p. 12). However, there were signs of tension between these countries in the summer of 1973. For one thing, the French were always treated with some reserve. It was clear that France had changed its policy toward the Arab world at the time of Algerian independence and looked for ways of translating the goodwill of this policy into some form of preferential treatment in oil matters. The Anglo-Saxons were not pleased when the French insisted on not giving any public sign that they were co-operating with other Western countries in establishing emergency oil procedures in 1967 (Church Report, 1975, pt 8, p. 549). In the same year, there had been considerable indignation within the industry and State Department when the French intervened in Iraq through CFP and ERAP in a way which suggested their policy was to undermine IPC's claim in its long-standing dispute with the Iraqi government (Church Report, 1975, pt 8, pp. 505, 546, 550*ff*.). In addition, there was strong suspicion that it was CFP which had leaked details to the Shah of the overlift formula of the Iranian Consortium in order to embarrass Aramco members (Church Report, 1974, pt 7, pp. 269–70). Both sides in the Teheran-Tripoli negotiations saw France as a potential blackleg and company representatives felt the need to keep CFP in line in order to disabuse the Libyans of the idea that they could count on the French (Church Report, 1974, pt 6, p. 31). At the same time, the Shah of Iran took comfort from statements by President Pompidou that, in the former's words, 'this situation cannot continue, that our [Western] money is devaluated, is inflated every year, and we sell you things more expensively every year and that we try to buy your goods at lesser prices all the time' (Church Report, 1974, pt 6, p. 162).

Ambivalent reactions to the French were reinforced in 1972, when the French government did a deal with the Iraqis that permitted CFP to continue to do business after the rest of IPC was nationalised. The other majors were quite happy to leave some channel open as a means of continuing negotiations on the various nationalisations, but there was unease lest the French were deviously gaining an advantage at the expense of others and, in particular, there was concern lest the principle of such bilateral, intergovernmental deals should become widespread. During 1973, the French further isolated themselves both from the USA and also from the EEC, which was slowly coming to grips with the need for some form of energy policy. The October 1972 Summit of the Community's heads of state included energy on the agenda, and in May 1973, there was the first ministerial meeting for over three years to discuss energy. The commissioner in charge of this issue, Henri Simonet, sought a mandate for talks scheduled later that month with the Americans and, the following month, within the OECD. The Eight were willing to push ahead with such consultations, but the French insisted that all negotiations should be carried out by individual member governments, and that the EEC should first work out a common energy position and unify its market before talking with outsiders. The resulting eighteen-hour wrangle effectively left the EEC unable to discuss energy problems with countries like the USA.

The French position appeared to stem only partly from oil interests. In the immediate aftermath of Kissinger's Atlantic Charter proposals, France seems to have reacted on the energy front in a way which was consistent with its general resistance to what the French saw as an American attempt to re-establish hegemony over Europe. However, the conflict over a unified EEC energy market reflected the ideological polarisation that existed between the French, with a tradition of tightly managed oil markets, and the British, Dutch and Americans, with their reliance on a free market of the oil majors. Whatever France's policies might be towards the Middle East, its *dirigiste* philosophy toward the oil industry was bound to make France a trying, unpredictable ally for the Anglo-Saxons, as events during and after the Arab oil embargo were to show (Turner, 1974, pp. 406–7). Even traditionally pro-American countries were not without some suspicion of US motives. There was, for instance, an awareness in Europe that the USA was becoming increasingly dependent on Middle Eastern imports, and there was a fear that the USA might somehow use its marketing strength and political clout to negotiate bilateral agreements with key Middle East sources, and thus pre-empt supplies which would otherwise have been bound for Europe, or Japan. Nervousness grew when, in the autumn of 1972, the Saudi government proposed a special oil relationship with the USA, by which Saudi oil would be exempt from restrictions and duties, and Saudi capital should be given increased downstream investment opportunities, thus offering hostages which would help guarantee security of supply (*Petroleum Press Service*, November 1972, p. 42). Although the US administration turned the offer down,

Europeans felt that the fact the approach had even been made was highly significant, and that there was an imminent danger of competition between Japan, the USA and Europe for Middle Eastern oil which could only lead to a bidding up of terms and prices.

There was also some US equivocation on the critical question of emergency allocation of oil supplies in future crises. The USA had warned the OECD oil committee in the late 1960s that in any future crisis it would no longer be able to divert domestic American production to European or Japanese markets. This was understandable but the fact that domestic production provided a large part of US oil needs meant that the USA would come through any supply disruption much better than those who were highly dependent on Middle Eastern oil. There were discussions within the OECD in 1972, on how to widen the emergency sharing scheme, but the USA refused to add its domestic production into the equation, taking the position that any sharing should be based on 'water-borne imports' rather than total energy requirements (Stobaugh, 1975, p. 185; Walton, 1976, p. 183).[10] As Stobaugh explains, such water-borne imports (which of course, ignored American imports from Canada) only accounted for 14 per cent of US energy supplies, compared with very much higher proportions for other OECD nations. A cut of 20 per cent in water-borne imports would reduce US energy supplies by around 3 per cent but Japanese ones by 20 per cent. The US administration sensed the need for some form of contingency planning but was not willing to sponsor a programme that would involve major sacrifices on the part of US citizens.

The Italians had long played a slightly maverick role in international oil affairs, with ENI's exclusion from the 1954 Iranian settlement a severe grievance which was more than compensated for by the irritation of the majors when ENI began purchasing Soviet oil in the late 1950s. There had been rumours in 1964 that Italy was trying to break into IPC's disputed territory in Iraq (Church Report, 1975, pt 8, p. 543). Despite hopes in the early days of the LPG that the Italians might be persuaded to join a common industry approach, this never happened. ENI went on developing its Libyan concessions through its subsidiary AGIP, apparently trusting that Italy's traditional ex-colonial relationship with Libya would see them through. Instead, AGIP was forced to conclude a participation agreement in September 1972 and was not, of course, eligible to apply to the Libyan Producers' Agreement for assistance.

Italian motives during this period are by no means clear, but the refusal to ally with parent governments was consistent with past policies. ENI, for instance, had first floated the idea within the EEC of 'Community oil corporations' to challenge the dominance of the non-EEC Anglo-Saxon majors. Although this idea was never implemented, by 1967 there was, according to two Italian observers, a loose agreement between the Italian, French and German companies (Prodi and Clô, 1975, pp. 105–6) which apparently took the form of attempting to do joint exploration work in Iraq.

The attempt to find a non-Anglo-Saxon approach to the oil industry culminated in the creation, over 1971–2, of a group of companies known as the 'Zurich group'. Very much Paul Frankel's brainchild, this was an attempt to 'de-colonialise' the oil industry. Frankel started in 1971 to get together companies from Japan, Germany, Spain, Italy and Austria and the following year Deminex, ENI, Hispanoil, OMV and JPDC began meetings. The aim of the Zurich group was to buy into existing concessions and a memorandum of understanding was agreed whereby no member would enter into an existing concern without prior agreement from the rest. There were some signs that the Japanese, who were particularly active in this group, were restive under these restrictions, but the events of late 1973 overtook the group whose last meeting was in early 1974.[11]

One of the most radical ideas to come out of Italy was the plan for oil companies to be turned into regulated utilities. The idea was that consumer governments would set prices that companies would pay for crude oil and charge for oil products, with the governments guaranteeing them a fair return on their investment. As Akins commented in 1973, this idea naturally horrified the majors and did not please him, but it did win adherents in consumer countries. It was extremely unlikely, in the circumstances, that the Italians would line up with Anglo-Saxon governments to defend the majors in anything but the most serious crisis.

At the other extreme was West Germany, which, although less well represented in the international oil industry, was not as automatically anti-American and anti-the-majors. Since the mid-1960s it had been taking steps to try to create a significant presence alongside the majors, but this policy seemed to be a precaution and not the result of any particularly deep hostility to the non-German companies which dominated its oil market. It is thus not surprising that Gelsenberg, the German company involved in Libya, should have sided with the majority of the industry through the LPG and the Libyan Producers' Association. What was in question was the extent to which German public opinion and political leadership was aware of the issue for which Gelsenberg was fighting. Some doubted that they would accept a confrontation between the oil companies and the producer governments if this would involve suffering for the German people – and any serious dispute with Libya would inevitably hit Germany particularly hard.

Japanese circumstances had some similarities with those of Germany. Both countries had pro-American leaderships and a tradition of tolerance to private enterprise (though Japanese attitudes to the majors were less favourable than those of Germany). However the Japanese are extremely dependent on imported energy and have had a far more active history of worrying about the security of their resource supplies than the Germans. On the other hand, as a European nation, Germany would instinctively be more wary of over-reliance on Middle Eastern oil, especially in the context of the vulnerability of the Suez Canal. The actual and potential threat to supplies from the 1956 and 1967 Suez crises made a greater impact in Europe than Japan. Germany's presence in the EEC and the OEEC/OECD meant that it was drawn into contingency planning from the beginning. Japan was admitted relatively late into the OECD and was not involved in the emergency allocation scheme concerned with supplies for Europe rather than for the total OECD membership. Japan did not have a stock piling policy in October 1973 although the Japanese had seen the need for a common approach towards OPEC which their Overall Energy Council had recommended in 1971, and AOC, their largest oil-producing company, was a member of the LPG. Moreover, Japan alone possessed a government department, the Ministry of International Trade and Industry (MITI), which could instruct companies not to offer exorbitant prices in Middle Eastern deals as the scramble for oil intensified during 1973 (*Petroleum Press Service*, October 1973, p. 372). The eight-day trip round the Middle East in May 1973 by the head of MITI indicated that Japan was no longer willing to rely entirely on the majors. The Japanese government by then believed that there should be strong and continuous oil diplomacy in which Japan, as one of the world's strongest industrial powers, would offer a co-operative relationship with the major oil producers. Offers to assist the promotion of industrial development were made in Iran, Saudi Arabia, Kuwait and Abu Dhabi, and were the kind of government intervention which was destined to worry the majors and their parent governments. The minister, Mr Nakasone, went out of his way to emphasise that Japan would not be taking any part in the oil importers' 'common front' which was being talked about in the USA and Europe. His position was indicative of the political distance which Japan was trying to put between itself and the parent governments in 1973 as it contemplated its extreme dependence on imported, primarily Middle Eastern oil (*Petroleum Press Service*, June 1973, p. 228).

The Arab embargo caught the importing nations generally ill-prepared, though they were aware that the world was moving into a critical era. The tightness of world markets was putting an upward pressure on prices, the role of the traditional companies in the Middle East was under continuous attack, respected foreign affairs journals were carrying weighty articles about the impending energy crisis (Adelman, 1972–3; Akins, 1973) and President Nixon's energy message to Congress in April 1973 announced that the US mandatory quota scheme was to be phased out over a period of years. An era was coming to an end. Consumers had hitherto been able to satisfy their oil needs without indulging in excessive competition. Now there would be competition for the oil from the crucial Middle Eastern area and oil imports which had been primarily a commercial affair were due to become a political issue in a struggle between consumer governments. In that kind of competition, the USA could be expected to win.

As 1973 continued a feeling of something approaching panic spread within the industrialised importing world. This was in the aftermath of the 1972 publication, *Limits to Growth*, by the Club of Rome (Meadows *et al.*, 1972), which had popularised the idea that the world was running out of critical raw materials, including oil. There was a temptation to interpret the global commodity boom of 1973 and the existence of oil shortages within the USA as symptoms of a long-term irreversible crisis, rather than as an inevitable result of a major world economic boom and resultant bottlenecks, made worse by tardy political reactions within the USA. BP's exploration manager argued that world oil production would reach its peak in the early 1980s and that, on then-current projected demand growth rates, demand could outstrip production as early as 1978. Overshadowing everything were the projections for US imports which, in Chase Manhattan's prediction, could well rise from 4 million b/d in the early 1970s to 20 million b/d in 1985. Even if the claims that the world was running out of energy could be dismissed, the danger that the USA might monopolise much of the world's production seemed very real (*Petroleum Press Service*, January 1973, p. 6, October 1973, p. 362).

However, reactions to the potential crisis were tentative and generally unco-ordinated at the international level. There was a plethora of studies and policy statements on the energy scene at national levels, particularly in the USA, but also in Japan, Canada and West Germany (*Petroleum Press Service*, February 1973, p. 48, July 1973, p. 267, August 1973, p. 308, October 1973, pp. 363–5, 367–9). Few of these had resulted in any policy decisions before the October crisis. The USA decided to phase out import quotas, and Nixon asked Congress to reform US energy policy to prevent existing national fuel shortages from becoming worse (*Petroleum Press Service*, May 1973, pp. 164–6). The Canadians took steps to increase their energy independence from the USA, imposing an export levy on crude oil destined for US refineries and deciding in principle to back the building of an extended pipeline system to directly link Albertan crude oil with markets in Eastern Canada such as those of Montreal and Quebec (*Petroleum Press Service*, July 1973, p. 267, August 1973, p. 308, October 1973, pp. 363–5). The West German government published a long-term energy programme concerned with issues like increasing oil stockpiles, consolidating the German-owned sector of the industry and including an attempt to integrate forecasts of developments desired in all energy sectors including coal and electricity as well as oil and gas (*Petroleum Press Service*, October 1973, pp. 367–9). Aside from France and its national plans, few non-communist governments had systematically looked at the energy scene in its entirety at this time and the German government was well ahead of its Italian, Dutch,[12] American and Japanese counterparts in its thinking (perhaps, also, of the British).

As the year progressed, there were some calls for improved intergovernmental action. Two of the more influential ones came from Jim Akins in his *Foreign Affairs* article of April 1973 and from the veteran oil consultant Walter Levy in a paper delivered at the Europe-America conference in Amsterdam at the end of March 1973. The latter called for an Atlantic-Japanese energy policy to be coordinated by an International Energy Council which was to do most of the things that the International Energy Agency created a year later eventually came to do – i.e. to encourage the development of added supplies of energy, diversification and research as well as to mastermind joint policies on stockpiling, rationing and emergency import sharing. The companies would then no longer be in danger of being singled out by militant governments as they would negotiate along the broad guidelines recommended by this agency. Needless to say, OPIC (Organisation of Petroleum Importing Countries) was never created, though Sheikh Yamani did find it necessary to warn the West that such an organisation would harm, not help, the interests of the consumers. Some of its functions could have been taken on by a more powerful OECD, but that body could then do little more than set up a group to discuss emergency energy policy in June 1973 instructed to report in November (Sampson, 1975, p. 242). Moreover, the refusal of France to allow the European Community to enter discussions as a unified group with the Americans meant that progress toward a consumers' front was effectively blocked. It did not help that the Japanese were nervous and were trying to see what kind of advantages they could gain through their own diplomatic initiative in the Middle East.

Thus the industrialised consumers faced the Arab embargo in disarray. There was a framework of defensive collaboration among the oil companies, but this was vulnerable to the vagaries of the US antitrust authorities and not designed to withstand a political onslaught. The emergency allocation scheme under OECD auspices had not yet been adjusted to the realities of Western hemispheric shortfalls. Moreover, this scheme required a unanimous vote by OECD members to be activated, and there were sufficient signs of strain within the Atlantic Alliance to make it doubtful if that unanimity existed. The general impression is that awareness and understanding of the impending crisis were still limited to the managers within, and diplomats closest to, the industry. In September 1973, after Libya had decreed a 51 per cent participation in the majors remaining in Libya at that time, President Nixon replied in a news conference:

> *Question:* What exactly are you doing to meet these threats from the Arab countries to use oil as a club to force a change in our Middle East policy?
> *Answer:* The radical elements that presently seem to be on ascendancy in various countries in the Mid-East, like Libya. Those elements, of course, we are not in a position to control, although we may be in a position to influence them, influence them for this reason: oil without a market, as Mr Mossadegh learned many, many years ago, does not do a country much good. We and Europe are the market and I think that the responsible Arab leaders will see to it that if they continue to up the price, if they continue to expropriate, if they do expropriate without fair compensation, the inevitable result is that they will lose their markets, and other sources will be developed. (Church Report, 1975, pp. 138–9)

To be charitable, President Nixon had other matters like Watergate upon his mind. His Mussadiq analogy had been valueless from at least as early as the heady days of 1970–1 when OPEC members had found their solidarity could help them defeat the consumer nations. His blunt distinction between 'radical' and 'responsible' leaders was also pointless – particularly at a time when the most 'reasonable' Arab leader of them all, King Faisal, was desperately sending messages through the American oil companies stating that he would be forced to take part in a general oil embargo against the United States if that country did not modify its attitudes on the Israel issue. Nixon was a fine example of Keynes's practical men enslaved by defunct economists. In this case, he was the quintessential pupil of virtually any Western oil industry economist at work during the 1960s, right down to his reiteration of the outmoded Iranian precedent. Mussadiq's fall had indeed inhibited host governments during the 1960s, but they had learned just how irrelevant that case was for the 1970s. The President of the United States, however, had not learned this and hence blindly led the Western world into the events of the autumn of 1973.

NOTES

1 Some company geologists, though, became increasingly pessimistic after the early 1960s.
2 The IIAB produced three reports during the 1967 crisis.
3 Odell has pointed out to me the significance of Venezuela during this period. It had, for instance, boosted its tax rate on the posted price to over 65 per cent by the autumn of 1970.
4 Membership varied slightly over time. For signatories to the original joint message to OPEC and the membership of the LPG and LPA, see Church Report (1974, pp. 234–5, 246, 247, 253–5, 263, 272).
5 Herbert Hoover, Jr had found himself thus constrained during the 1951–4 Iranian crisis (Church Report, 1974, *IPC*, p. 49).
6 For biographical details see Church Report (1974, pt 5, pp. 59–60).
7 See also the evidence of Akins, Irwin, Schuler, Piercy and McCloy in Church Report (1974, pt 5).
8 Rustow has informed me that the Libyans had foreign exchange reserves for three years of imports as against Europe's approximate sixty days of crude oil in reserve.
9 There is some ambiguity in the reports. See Church Report (1974, pt 6, p. 12).
10 It has been suggested to me that the USA would have shifted from this position if necessary.
11 Personal communication from Paul Frankel.
12 By 1973 oil was of minor importance to the Dutch energy economy.

9
The Embargo and After – Testing Time

The handling of the Arab oil embargo announced in October 1973 was a short-term triumph for the traditional oil companies. In retrospect, though, the events sparked off by that crisis can be seen as a turning point in the relationship of the companies with the industrialised world as significant as the Teheran-Tripoli Agreements proved to be in their relationship with host governments. In the short run, they moved purposefully and effectively into the policy vacuum of their parent governments. In the long run, the issues raised during those testing months were such that calls for governmental intervention by parent and industrialised consumer governments were inevitably made and listened to. Whatever freedom of action the majors may have had in the past from such authorities was lost. As after Teheran–Tripoli host governments encroached on the freedom the companies had won for themselves in the late 1950s and 1960s, so, in the aftermath of the embargo, OECD governments realised that oil was too important to be left to the oil companies alone. If the latter had occasionally come close to being untrammelled transnational actors, those days have now gone for the immediate future.

CONSUMER GOVERNMENT DISARRAY

There were several levels at which the OECD nations showed disarray in the face of the Arab embargo. First, there was the OECD's specific decision not to activate the European emergency allocation scheme. Secondly, European governments were divided amongst themselves on the question of how they should respond to the Arab decision to penalise one of their smaller members, the Netherlands, while apparently giving preferential treatment to two of the larger countries, Britain and France. Finally, although there was a certain amount of lip-service paid to the idea that the OECD nations should not make a difficult situation worse by bidding against each other for the limited supplies of crude oil, in practice a number of them were quick to send diplomatic-cum-commercial delegations round the Middle East with the prime purpose of signing preferential deals for the future supply of oil. The companies were apparently given minimal and in some cases contradictory, political guidance on how to handle the embargo and they viewed the politically motivated direct deals for oil as a dangerous long-term threat to their traditional role.

On the surface, the absence of any decision of the OECD oil committee to activate the emergency allocation scheme appears as an act of cowardice by governments unwilling to align themselves publicly with aggrieved nations such as the USA and the Netherlands. There were, however, good reasons for this policy. There was no point in aggravating the Arabs into making severer general cuts in production which would hit everyone, and there was also a clear understanding within OECD circles that the oil companies would be sharing the oil around. But the dominant reason, perhaps, why the scheme was not activated and the International Industry Advisory Body brought into action was that, thanks to consumer government inaction in the previous years, it was the wrong scheme for this particular crisis. It was a European scheme and thus inadequate to a situation where the USA was a prime target and oil users throughout the world were affected by the Arab decision to restrict supplies over and above embargoes against specific countries. The last crisis in which the OECD, then the OEEC, had been called in to play a major role had been in 1956, and that had been a simple situation compared to 1973. As Ulf Lantzke, the first executive-director of the International Energy Agency, explained, fear of offending the Arabs may have been one factor in restraining OECD members, but the really decisive element was the lack of information, and the feeling that the companies alone possessed the logistic systems needed to handle such a fast-moving and complex crisis. The OECD oil committee came to play a supportive role. It kept an informal check on what was happening, acted as a clearing house for key statistics like up-to-date stock figures from various countries and handled certain aspects of the crisis which could not be dealt with on a national basis, such as marine and aviation bunkers (Lantzke, 1975, pp. 219–20). So there was a political vacuum within the OECD and the crisis moved too fast for there to be any serious attempt to link the logistic systems of the various companies in any meaningful way though systematic information gathering would have helped. Without that, there was no point in trying to put together a joint approach by the companies to the supply problems. Each of the companies made its own choice as to how it would share the shortages and inter-company co-ordination seems to have been carried out on a discreet basis, presumably by the OECD oil committee, which kept up consultations with each of the companies, though not consistently or concurrently. There was some exchange of views about the way supplies were moving, and even though this was obviously a rough-and-ready way of handling the situation, it seems to have worked (Lantzke, 1975, p. 219).

There were, however, at least two drawbacks to this very low-key approach to industry-government co-ordination. The first was that although those actively connected with the OECD oil committee might have a pretty clear idea of what was happening, many national leaders did not, and some of these did not subscribe to the idea that oil shortages should be spread around. The Tory government in Britain, led by Edward Heath, put heavy pressure on BP and Shell to maintain their supplies to Britain as a first priority. There was a meeting on 21 October between Edward Heath, Frank McFadzean of Shell and Sir Eric Drake of BP. The latter insisted on the belief, founded on legal advice, that discrimination between contractual customers would lead to heavy law-suits against them, perhaps even expropriation. If the British government really wanted more than their share of available oil, then they would have to pass the necessary legislation to allow the companies to declare *force majeure* (Sampson, 1975, p. 263). The French were more formal in their insistence that they should be given the preferential treatment to which their designation by Saudi Arabia as a priority nation entitled them. They not only knew what oil deliveries were scheduled from which source before the embargo, but instructed the companies to complete all deliveries and to deliver as well the extra Saudi Arabian oil to which France was entitled as a preferred nation. Italy asked the companies to continue full, normal deliveries of crude oil and both Italy and Spain declared restrictions on oil exports (Stobaugh, 1975, pp. 189–91). According to the evidence, the companies resisted such pressures; there are signs that the British and the French both did better than they might otherwise have expected, although a statistical comparison between the three months of the embargo with the equivalent three months of the preceding year is inconclusive (FEA, 1975b, pp. 19–30).

However, whether or not the British, French and Japanese did come off slightly better than other countries which pressured the companies less, it is clear that the conflicting pressures on the companies were not as strong as might appear on the surface. In Britain, it seems to have been Edward Heath and one or two of his Cabinet colleagues who called for a 'Britain first' line. Within the Foreign Office there were critics of this view and the companies did not lack sympathy among informed opinion for the conflicting pressures put on them. According to Stobaugh, CFP followed the same allocation policies as the other majors (1975, p. 190), despite its French government stake, and there is a feeling that the tough oil policy of the then foreign minister, Jobert, was not universally supported by French political and administrative elites. In all countries, the companies had influential supporters who from the start knew and approved of what the companies were trying to do. Kissinger, for instance, held three meetings with US oil company executives between October and December 1973 in which he was briefed and apparently indicated that the Netherlands should be taken care of and the Japanese treated fairly (Church Report, 1974, pt 7, pp. 450–1; Stobaugh, 1975, p. 188). With this kind of backing, the disarray of importing countries was not too serious.

However, a second disadvantage of the low-key, informal way in which the companies handled the crisis was that most governments, media and public opinion had no clear idea of what the companies were doing. This meant that the situation was ripe for rumour. Stories multiplied of tankers waiting off the coast of the USA to take advantage of price hikes, and of supplies diverted within Europe to countries without rigorous price controls. Of all the countries concerned, Japan seems to have been the most unsure of what was happening, with its domestic refining industry predicting that crude oil arrivals in January 1974 would be 50–70 per cent below normal and, for a while, convinced that the American companies were diverting oil from Japan to satisfy the embargoed US market. Confusion was not a Japanese prerogative and when Kissinger visited Japan in November 1973 he was apparently shocked not only by the depth of Japanese suspicions but also by the fact that the State Department was not in a position to give authoritative reassurances. As a result, he ordered a survey through the relevant US embassies of what was happening. The fact that such a survey was necessary convinced him that some form of governmentally approved emergency scheme must be erected before the next crisis (Stobaugh, 1975, p. 192). The idea that Walter Levy had floated earlier about the need for some form of international energy council re-emerged in Kissinger's Pilgrims' Dinner speech in London on 12 December 1973, when he called for the creation of an energy action group charged with developing an international action programme in response to the embargo and the concomitant price increases (Walton, 1976, pp. 183–4).[1]

During the period when this idea was being transformed into the International Energy Agency, there was considerable activity by consumer governments seeking to sign bilateral agreements with individual OPEC governments in the hope of winning preferential treatment for oil supplies. The Japanese were notably quick off the mark. Having got themselves off the initial list of countries 'unfriendly' to the Arabs (and incidentally thus making one of their first major postwar breaks with US foreign policy), they sent a series of special envoys, beginning in December 1973, to visit Arab nations. They offered aid and technical assistance in profusion and, by mid-1974, had promised some $563 million to Iran, Syria, Egypt, Saudi Arabia, Algeria, Sudan, Jordan and Morocco (Tsurumi, 1975, p. 124). The French signed a series of deals with Saudi Arabia, Libya and Iran (the rest of the Western world was particularly unnerved by a potential twenty-year deal with Saudi Arabia which was to have covered nearly 5 billion barrels of oil). The British and the Italians signed deals with Iran. As 1974 progressed, the Germans made progress toward an understanding with the Iranians involving some thirty-five long-term projects. The USA received Prince Fahd and Sheikh Yamani in Washington, reviving European worries about a Grand Saudi-American Bilateral Deal which would squeeze the rest of the world out of the market for Saudi oil (Sampson, 1975, p. 276; Mendershausen, 1976, p. 75; *Petroleum Press Service*, March 1974, pp. 82–3).

The enthusiasm amongst consumer governments for this kind of deal waned in 1975. By June 1976, the French were quietly discouraging the Iranians who were proposing to barter crude oil for goods and services, the former having learned from their one small deal with the Saudis that bilateral deals do not guarantee low prices, and from other lengthy negotiations that significant diplomatic effort does not necessarily mean that any deal gets finally signed (*Petroleum Intelligence Weekly*, 7 June 1974). On the other hand, the US government was investigating in mid-1976 the possibility of such intergovernmental deals on the supply of crude oil to the new US national stockpile and was also investigating some form of arms-for-oil barter with Iran. In general, though, consumer governments were wary of direct deals once the initial panic of the various importing nations died down and the realisation grew that some form of united approach to the energy issue might be necessary. That this mood was sensed and channelled to create the International Energy Agency is very much a tribute to the positive diplomacy of Kissinger who drew on American experience to convince the Europeans and Japanese that there was indeed an alternative to beggar-my-neighbour bilateral oil deals. He was not always very sensitive about how he went about approaching his goal (Turner, 1974b), nor were his proposals always realistic. It is said that this was the first time he seriously had to come to grips with economic issues (though in November 1971 he had to wrestle with the international implications of the dollar collapse) and his inexperience sometimes showed by overstressing issues like the $7-a-barrel floor-price. He did contribute the conviction that something had to be done fast, that decisions had to be taken and momentum had to be maintained in building up an institution like the IEA into an effective body. Although some of his ultimatums seemed unnecessary (such as insisting that the USA would not allow the north-south dialogue to get off the ground until IEA members accepted the principle of a floor-price), he was generally willing to give ground when it did become clear that his position had become untenable. He originally intended to keep the IEA outside the OECD, because of the latter's unanimity principle, but was willing, in the hopes of attracting France, to accept a compromise whereby the IEA would operate within an OECD framework, with a carefully devised form of majority voting. His stubbornness at times infuriated others but ensured that his proposals were put into practice with just enough of a compromise to satisfy the Europeans and Japan that the IEA would be working in the interests of all, and not just of the USA.

With the exception of the hostile French, it is clear that Kissinger was pushing at an open door and that the OECD nations (though sometimes a little concerned lest the French were correct in asserting that relations with the Arabs could be damaged by entering into negotiations with Kissinger on this issue) in fact shared most of his worries. The British and Dutch certainly realised the need to improve consultative procedures in advance of new crises. The Germans were motivated by their traditional desire to maintain the credibility of the Atlantic Alliance, but inevitably came out of this crisis, in which their oil supplies depended so heavily on the relationship between the Dutch and the Arabs (since German oil predominantly flows through Rotterdam), supporting all realistic proposals for reducing the impact of future crises. However, the fact that governments like those of Italy and Japan, which had ambivalent roles in the period leading up to the embargo, were willing to back the IEA, indicates that the bulk of the industrialised world really did feel a mutual solidarity in the face of the Arab embargo and OPEC price rises. They accepted the need for an intergovernmental body charged with devising an effective emergency allocation scheme and master-minding the international aspects of the search for new forms of energy production and conservation. What Kissinger did was to provide the leadership which brought out the potential co-operation that already existed. The diplomacy which started in December 1973 with his call for consumer co-operation evolved through the Washington energy conference of February 1974 and the temporary body known as the Energy Co-ordinating Group to become in November 1974 the International Energy Agency, an offshoot of the OECD (Walton, 1976).

The IEA has been a major institutional innovation. It is true that the old OECD oil and energy committees tried to perform some of the tasks that the IEA has set itself, but there is no strict comparison. The former emergency allocation scheme was a rough-and-ready European affair, while the IEA's scheme is on a different level of technological sophistication, embracing most the major non-communist nations with the notable exception of France.

In the field of oil and energy policies, the OECD committees were rarely allowed to pass judgement on individual national policies. The IEA is instructed to co-ordinate a wide range of international research programmes which would have seemed unbelievably ambitious before the embargo. As governments have moved more formally into the energy field, the role of the international companies has altered. On balance, the countries involved in the IEA see the organisation as a means to complement, rather than supersede, the companies. The companies themselves have a formal supportive relationship to the IEA accepted by all members, in which they have to give a great deal of information about corporate activities which could, theoretically, be used to help the fortunes of smaller, national competitors.

The new arrangement, however, seems unlikely to affect the freedom of the companies to negotiate with host and producer governments. For instance, the IEA is not going to pass judgement on whatever deal Aramco and the Saudis may come up with. Where it may come to play a significant role in the long run is through fora like the north–south dialogue (the Conference on International Economic Co-operation) where it was charged with formulating the consumer response to the Third World group within the Energy Commission. Since issues like the indexation of the price of oil to the price of industrial goods were being discussed in this forum, it was possible that OPEC decisions about the general level of oil prices might come to be affected as much by CIEC as by any representations from the oil companies. However, there is no evidence at present that the IEA will give instructions on the limits within which to conduct oil purchases. Companies have the freedom to accept or reject any particular bargain which is offered to them. With the full knowledge of all member governments, the companies were given a formal advisory status to the IEA from the start. In the beginning there were two active industry working groups (one chaired by BP, and one by Exxon) which advised the IEA on its emergency sharing scheme and forecasts of world oil markets. The US authorities gave antitrust clearance first through a Business Review letter, then through the Defense Production Act and finally through provisions in the Energy Policy and Conservation Act of 1975. The events during the embargo also alerted EEC antitrust authorities to the potential implications of industry collaboration in emergency allocation, and its Competition Directorate began to send observers to industry group meetings

(*Petroleum Intelligence Weekly*, 8 March 1976, p. 12, 10 May 1976, p. 4).

Early in 1975 there was some discussion within the OECD about setting up a permanent industry group to sit with the IEA in discussions of major issues, but this idea seems to have lapsed. The IEA now works with two advisory industry groupings, one now known as the Industry Advisory Board, the other known as the Industry Working Party. (The IAB is concerned with emergency allocation issues and consists of sixteen companies, of which seven are neither British, American nor Dutch, but probably include Petrofina, the Austrian OMV, Deminex or Veba-Gelsenberg from Germany, ENI and a Japanese representative.) These groupings (the membership is slightly different) are a further step in what may become the evolution of the companies into accredited agents of the industrialised importing countries. What is particularly interesting is that the range of companies involved has once again been broadened to include more from second-ranking oil powers, thus giving the latter a greater involvement in oil councils and their national companies a more formal recognition than they have had before.

There are two ways in which the IEA is supporting the position of the companies as general agents for consumer governments. First, the formulation of a substantive emergency allocation scheme is a useful device to strengthen the industry's bargaining hand. It is not intended to have the same function as the Libyan Producers' Agreement, that is, to protect individual companies in dispute with particular countries, although there seems no reason why a safety-net could not be re-created if needed. What it does is to show OPEC that a disruption of supplies in the course of a commercial or political dispute may now be met by a united consumer stand. This is a major change from the 1970–3 period when OPEC could be pretty certain of disarray within the industrialised world. At the same time, the fact that a formal structure exists to come into action when oil-flows are cut means that the ambiguity about the companies' role in a crisis will be removed. Consumer governments will be seen to be masterminding the industry's overall reaction, so that government and popular suspicion of the companies' actions should be avoided. But should there be another embargo crisis, OPEC control will undoubtedly be more specific and better co-ordinated than in 1973 when only the Saudis and Kuwaitis gave detailed instructions. In the future more Middle Eastern oil will be travelling in Arab ships and thus responsive to an Arab embargo.[2] Since the IEA is trying to extend its authority over oil-flows in the interests of the industrialised nations, the chances are that the companies may have to square even more incompatible instructions in a future crisis. The companies will presumably obey OPEC instructions as far as the initial destinations of the relevant oil is concerned, but would then feel free to trans-ship these deliveries wherever they choose. But the relatively informal monitoring methods used in 1973 have presumably improved to reduce the area in which sleight of hand by the companies could be used to reallocate oil-flows. However, should OPEC try to stop the companies from substituting non-embargoed for embargoed oil, the IEA would then give balancing instructions which would presumably make the consumer governments, rather than the companies, responsible.

The second way the IEA is supporting a new role by the companies – ex-majors, independents and state companies – is by its adoption of a minimum safeguard price, or floor-price. The politics behind the IEA's acceptance of a $7-a-barrel floor-price were complicated and do not concern us here (Walton, 1976, pp. 192–6). The need for such a device arose from the fact that the price set by OPEC is so far in excess of actual production costs that it could at any moment fall steeply as a result either of OPEC's break-up (unlikely) or of a conscious decision to sabotage the search for alternative sources of oil and energy. Implicit in IEA acceptance of the floor-price principle is an acceptance that the private or semi-private corporations which are going to be involved in this search will not be able to invest in the development of high-cost energy unless they can reassure the international financial community that money invested in such projects will not be thrown away should OPEC members for any reason, decide to lower the price of their oil. In this way the IEA has pioneered a form of industry-company relationship which is unusual by the standards of latter-day capitalism (outside the agricultural sector). Member governments are choosing to reduce the risk for companies developing alternative sources of energy, thus allowing them to consider projects which would otherwise be rejected as too speculative.

The price for this wider legitimation of the companies' role has been that they must accept a far greater degree of 'transparency' in their operations than in the past. This trade-off was inevitable afer the embargo period brought home to a wide range of governments how little they knew about how the oil companies worked, and how few means they had to improve their knowledge. Even if a government was able to collect reliable information it had no way of comparing its experience with other governments. For instance, national tax authorities have had difficulties with the industry's transfer pricing policies. (Just how much should a company be charging for transporting crude oil? What was the market price for crude oil and its products at any one time or place?) The question of market prices has always been particularly difficult since the volumes of crude oil which pass through 'non-integrated' hands has traditionally been well under 10 per cent of the total volume being moved. The open market exchange of crude oil and products has generally developed round the larger concentrations of refineries in places like Rotterdam (hence 'the Rotterdam market') and the Caribbean. It has operated at the margins of the industry, being a market through which large companies can dispose of temporarily surplus crude oil, or where refiners can offer products for which they have no long-term contracts. Because it is a marginal operation, the prices at which oil is traded are always subject to wide fluctuations. In times of glut all companies try to get rid of the last 5 or 10 per cent of their products and find few buyers; in a seller's markets everyone – majors and independents alike – enters the market looking for the extra few percentage points of supply which will allow them to capitalise fully on surging demand. The prices established at such times are not representative of the prices at which the bulk of oil, passing hands under contracts of various sorts, is actually transferred. In times of gluts they will be much lower, and in times of shortage, much higher. Government officials have frequently been faced with the thankless task of trying to decide the extent to which the price involved in such a spot-market transaction should be used as a benchmark, or whether the transfer price which an oil company sets for one country is comparable to what the same company is charging neighbouring countries.

Such issues have not worried parent governments unduly, since they have assumed that overcharging by the companies would result in remitted profits which they can tax. Non-parent governments, such as the Japanese or Italians, have obviously been much less sanguine and they insisted in the bargaining which preceded the creation of the IEA that parent governments ensure that the activities of the oil companies must be more open to inspection. The Italians seem to have been particularly keen in their insistence but even German voices, normally staunch defenders of free enterprise, were raised in sympathy.

The form this public scrutiny is taking is, without doubt, something unprecedented. The IEA performs a quarterly analysis of the prices of sixteen key crude oils which make up over 80 per cent of the world's trade in crude oil. The raw data is collected and aggregated by IEA member governments; the aggregates are brought together by the IEA and then fed back to the national authorities. In mid-1976 the statistics covered the loading price of crude oil at the country of origin, sometimes the landed price at the importing end and the actual 'prime acquisition cost' incurred by companies. There are plans to get forward estimates of the capital demands of the oil and gas sectors, the aim being to see the extent to which existing companies are investing in future resources and what the economic and financial constraints are on such investment.

Not surprisingly, the companies have been somewhat nervous about the whole exercise. The majors are only too ready to believe that a country like Italy which has a record of giving considerable help to its national company could easily take IEA information and use it to help ENI strengthen its position in world markets. Hence the Americans, with the support of the Dutch and British, fought a battle to ensure that the information presented would be as anonymous as possible. The companies were not too worried that the Italians, say, would know that there was a significant discrepancy in the price being charged for a given Libyan crude oil between Italy and Germany. What they wanted to avoid was a situation in which the Italians would know the exact terms under which a named company had acquired, transported and disposed of a given crude oil in any part of the world, since this information, if passed to ENI, would allow the latter to know to a nicety what prices to charge or bid in order to undercut. In the end, the IEA's scheme involves safeguards to protect the anonymity of individual transactions. No price is quoted if an importing country has only one firm providing a particular crude oil in any quarter, and there are some special safeguards if only two companies are involved. There is even more security about acquisition costs; these are fed by government couriers straight into IEA computers and then the raw data are removed. The computer memory is erased the minute the aggregated results have been produced, so that no-one will be able to tap into the computer and find out exactly what a specific company is paying for any given crude oil (*Petroleum Intelligence Weekly*, 22 March 1976, pp. 7–8, 10 May 1976, p. 5).

Governments do seem to be finding this information of use. According to *Petroleum Intelligence Weekly*, there have been cases of consumer governments successfully pressuring importing companies into cutting prices. There has also been a noticeable narrowing in the range between the prices charged for individual crude oils at the importing end. The existence of this information system reduces the importance of one of the arguments for the creation of national champions, which it was claimed, would provide governments with a 'spy glass' into the inner workings of the industry. Although in an age of increasing economic nationalism the national champion approach will inevitably grow, the new 'transparency' increases the chances that a government would be willing to work with the local subsidiaries of the majors to further certain national goals.

POPULAR DISCONTENT

But while the companies were winning the tacit approval of consumer governments, they were not in popular favour, particularly in the United States. It is difficult to know what liberal interest group in the USA had not been offended by the oil industry by the time of the embargo. To the traditional critics of depletion were added the anti-Vietnam War protesters who became increasingly convinced after 1970 that the prospects of oil discoveries off the South Vietnamese coast were one reason why the war was allowed to drag on (Howell and Morrow, 1974, pp. 121–9; Sale, 1975, pp. 258–91), the environmentalists to whom the Santa Barbara oil spill of 1969 and the Alaska pipeline symbolised everything wrong with corporate attitudes on this issue, the Watergate investigators who could not ignore the fact that oil companies had been involved in making illegal contributions to the 1972 Nixon campaign, and Zionists who were deeply and vociferously offended by the industry's backing of the Arab cause, particularly Socal's letter to its shareholders and Mobil's advertisement of the summer of 1973 (Sampson, 1975, pp. 246–7). Finally there was the ordinary man-in-the-driving-seat who was puzzled, then increasingly incensed, by the gasoline shortages which occurred in the USA from 1971 onwards. In a period in which a Nader-like suspicion of Big Business was very pervasive, it was easy to see the original, pre-embargo shortages as a corporate plot to allow oil prices to rise – and, when the embargo was finally declared, to see the resultant extra shortages as confirmation of corporate wickedness. In the final days of the embargo many Americans believed that it was the companies who had been holding back supplies (McKie, 1975, p. 85). The story that audiences stood to cheer at the line 'Down with all tyrants! God damn Standard Oil' in a New York revival of Eugene O'Neill's play *A Moon for the Misbegotten* may be apocryphal, but it expresses a very definite public mood. It was not until the third quarter of 1974 that the companies stopped declaring substantially increased profits, and it was obviously a public relations nightmare trying to explain these away to gasoline-starved citizens. There was a flurry of Congressional investigations, the most intensive being Senator Church's Sub-Committee of the Senate Committee on Foreign Relations, which started hearings in early 1974 in the hopes of discovering a scandal equivalent to that of ITT's involvement in Chilean politics. Its final report in January 1975 was not notably hostile to the companies, though it did argue that the majors were now providing OPEC with important advantages: 'to maintain [their] favoured status, the international companies help proration production cutbacks among the OPEC members. Their ability to do this derives from the existence of their diversified production base in OPEC countries' (Church Report, 1975, p. 10).

It may well be that the conclusions were less important than the fact that the hearings turned the searchlight on the relationship between the companies and the State Department and on the tax decisions which seemingly gave the companies unusual favour. Revived interest in the antitrust case which had come out of the Federal Trade Commission's report in 1952 and doubts on the strategy the companies and the State Department had followed in 1971 left a suspicion that the companies and OPEC were working to each other's advantage. Anthony Sampson worked closely with the sub-committee's investigators in writing his best-seller, *The Seven Sisters*, which was to amplify their conclusions and perpetuate a picture of the international oil companies as too big and powerful for the public good. Company protests fell on deaf ears and John Blair, co-author of the FTC Report, came out with his own hostile analysis of the industry in his *Control of Oil* (1976).

In the circumstances, it was inevitable that the future of the oil companies should become a matter for public debate within the USA. At first this was an attack on anything in the administration's slowly evolving energy policy which looked as if it favoured the companies. The industry was stripped of depletion allowances on most crude oil and natural gas and also of much of the credits it had been allowed against US taxes on foreign income.[3] A more intractable problem was the question of price de-control and what should be done with the resultant 'windfall' profits. This was a dilemma faced throughout the world and was partly caused by the fact that new supplies of energy would have to be developed and produced by commercial companies guaranteed some form of 'fair' return on their investment. The United States was faced with the problem that allowing the prices in its large pre-existing domestic energy industry to rise to the inflated OPEC level would bestow an immediate 'windfall' profit to companies with energy sources developed in the era of low costs.

This was not just an American problem. The West Germans had to face it with their relatively small-scale domestic production and the British and Norwegians had to bring in special taxes to curb potential 'super profits' in the North Sea. However, what happened in the USA is of central importance. Initially, before the post-embargo price explosion, it imposed a two-tier price ceiling with 'old' oil to be sold at prices prevailing in March 1973, with a 35 cents per barrel addition, and 'new' oil (imports and oil produced in excess of 1972 levels) to be uncontrolled (McKie, 1975, p. 77). The Nixon/Ford administration was soon convinced the 'Project Independence' energy package it was trying to put together through 1974 and 1975 must entail de-regulating the general price of oil. Otherwise it would be trapped in schemes of byzantine complexity in which refiners with access to cheap 'old' oil were expected to compensate refiners who were dependent on 'new' oil (in the autumn of 1974 'old' oil was controlled at $5·25 per barrel – about half the price of imported oil) (*Petroleum Economist*, November 1974, p. 407). The general position of the Democratic majority in Congress was that, if anything, the price of oil should be reduced, and there was certainly little enthusiasm for boosting prices in the run-up to the election year. The resulting paralysis in US energy policy making led to the compromise of the Oil Policy and Energy Conservation Act of December 1975 which extended the administration's powers to control oil prices at least until 1979, reduced the average price of the mixture of 'old' and 'new' crude oils for 1976 and allowed the president to raise the ceiling by a maximum of 10 per cent per annum after then (*Petroleum Economist*, January 1976, p. 10).

Strangers to the complexities of US energy policy may find it difficult to assess the implications of such decisions. What they illustrate is the kind of battle which inevitably arises when a price regulation mechanism affects both consumer and producer, particularly when it is decided that the consumer should make short-run sacrifices in the long-term interest of higher production. If it is decided that the financing of future energy production will be carried out by private companies, then relaxing price controls to allow them to generate profits to be invested in further productive facilities is the simplest way of handling matters, which is what the Oil Policy and Energy Conservation Act set out to do. Manipulating price levels to influence future investment is an uncertain approach as companies may well re-invest profits in unrelated activities (as Mobil has done with the store company, Marcor). It can be politically explosive since it is extremely difficult to convince consumers that a higher price today will guarantee more secure or cheaper supplies of energy in five or ten years' time. Most other OECD nations have managed to handle this issue more easily as most of them have a more centralised and less transparent decision-making system than the USA, with its separation of power between president and Congress and extreme responsiveness of members of Congress to the short-term needs of their constituents. Another reason why the USA has had more difficulty in raising energy prices is that, of all the industrialised nations, it is a society built on cheap energy. Japan and most European nations have not developed land-use patterns which are so heavily dependent on the availability of plentiful supplies of cheap gasoline.

It would be ridiculous to imply that the split between the interests of the producer and consumer is confined to the USA. Companies are faced with price controls on oil in the majority of the industrialised countries and in recent years there have been a number of disputes, notably in Italy, on the right of companies to pass on OPEC price increases to consumers. The Canadians have long had to balance the interest of their western producing provinces and their eastern consuming counterparts. During 1975 there was a bitter dispute when Ontario extended a price freeze on oil products at the time when Federal policy was to raise all Canadian prices toward the international level (*Petroleum Economist*, October 1975, p. 376). Similar tensions have arisen within the EEC, where France has resisted the attempt of the North Sea oil and gas producing states to get a unified floor-price for European energy.

Within the USA distrust added a major complication to the search for long-term energy security in the form of the demand for the dismemberment (or 'divestiture') of the integrated companies. In a mood somewhat reminiscent of that in the years leading up to Standard Oil's dissolution in 1911, Congress debated a number of measures which would require either 'vertical' divestiture (a company would have to choose between producing, refining or marketing oil) or 'horizontal' divestiture (an oil company would not be allowed into the coal or nuclear fields).

Whether some form of divestiture requirement is ever passed by Congress or not, the fact that all countries are starting to take action, both nationally and internationally, to devise energy policies makes it seem extraordinarily bad timing to start dismembering the financially and managerially strong companies now. There are strong reasons to believe that divestiture would slow down the development of alternative energy sources upon which at least part of a well-thought-out scheme for national or international security must rest. Thus the debate may lead the USA into a vicious circle in which popular distrust of the companies results in a delay in decision making which then leads to shortages and higher prices which are then blamed on the companies in new, drawn-out debates. The belief that the public good will be increased by dismembering existing companies, rather than by searching for ways in which institutions representing the public interest might be given greater control, seems odd to Europeans. One reason why the former approach has been preferred is that it has represented the highest common factor round which all critics of the industry can be united. Any proposals involving increased public intervention in the running of such companies starts alienating important groups. The idea of nationalising them is far too radical for a country which desperately tries to keep private enterprises alive until they reach such dire straits that some form of nationalisation seems the only alternative to bankruptcy and massive public disruption (Penn Central).

However, President Carter's energy proposals of April 1977 have shown quite clearly that his administration is not enthusiastic about dismembering the companies, though he did specifically point to company diversification into coal as an area demanding special vigilance. Rather, he has chosen to push American policies much closer to the mainstream of practices elsewhere in the OECD world. The creation of a unified Energy Department should remove the anomaly of the most important energy-consuming nation in the world being also the one in which the formulation of relevant policies has been most noticeably split between numerous competing agencies. The decision to ask US consumers to pay world prices for their oil is very much a policy approved of by other OECD nations, which have been having to pay such prices themselves and have been increasingly annoyed that US cheap energy policies have only encouraged world consumption of oil, thus intensifying competition for this depleting resource. It is not clear, as I write, to what extent the traditional vested interests (the auto manufacturers, the oil companies, consumer lobbies, state governments, etc.) will emasculate his proposals, but it is clear that American policy is becoming less idiosyncratic than it has been over recent decades. The fact that the US government was also in April entering the world crude oil market directly for crude oil for its 1 billion barrel emergency stockpile was a further sign that there is room for an American National Oil Company to carry out certain tasks (such as this) which cannot be left to the traditional private companies for political reasons. If such a state company does emerge, then the USA will probably have gone as far along the path of state intervention in the oil industry as will be necessary to satisfy public opinion – and it will finally be in step with most of the other major industrialised countries.

The EEC antitrust authorities gave a qualified clearance to the companies after the embargo, stating that they were generally satisfied, but still wanted to investigate four areas in which some illicit manoeuvring might have taken place, including the supply of kerosene to airlines, naphtha to the chemical industry and fuel oil to electricity producers (*Petroleum Intelligence Weekly*, 5 January 1976, p. 3). There was no suggestion, however, that the companies had unfairly discriminated between countries within the EEC.

Within Europe, German concern about how it had been treated during the embargo manifested itself in proceedings by the Federal Cartel Office against Arab, BP, Shell, Texaco and Esso on charges that they had misused their market domination to keep the retail prices of gasoline artificially high. In April 1974 there were much-publicised hearings into allegations of overcharging on heating oil and diesel fuel, and the Federal government announced they would initiate an international investigation into oil company marketing and pricing (which explains, in part, the German interest in the transparency issue within the International Energy Agency). Companies like BP and AGIP took to the courts to dispute various Cartel Office rulings, and the general proceedings were dropped in the summer after a certain amount of information swapping between the Federal Cartel Office and US investigators such as the staff of the Church Sub-Committee (*Petroleum Economist*, September 1974, pp. 336–7). The French hardly needed any reminding of the potential dangers of doing business with the majors and there was growing popular unease that the two main French companies, CFP and Elf-ERAP, were little better than the majors. There was a concerted industry campaign during the summer of 1974 against too-restrictive price controls, and this united the foreign companies, M. Guillaumat of Elf and even the government oil agency DICA (Direction des Carburants) against the Finance Ministry which determined price levels. Popular feeling was particularly stirred by the allegations of the Schvartz Report of a special parliamentary commission of inquiry in November 1974 which criticised various forms of 'anti-social behaviour' on the part of the companies, particularly focusing on the two national companies and their ambiguous relationship with the French government. It was a report which had been done on the initiative of the Communist Party, though M. Schvartz was a Gaullist deputy, and thus represented a wide-ranging attack from both ends of the political spectrum against the alliance of private oil interests and political administrators which had governed French oil policy for so long (Mendershausen, 1976, pp. 95–6). In the background, there was a long-running legal case in Marseilles in which a small independent businessman claimed that the larger companies doing business in France had tried to drive him out of business. Well publicised, this was a case which would come to public attention at regular intervals, as in mid-1976, when it was alleged that the magistrate involved was deliberately being moved to another district for political reasons. Given the polarisation of the French political system, one should not play down the importance of such *causes célèbres*. Distrust of the oil industry runs deep and should the Communist Party come to power in France, a complete restructuring of the industry, including nationalisation of the private holdings in CFP, seems likely.

Italian hostility to the industry was strengthened when, in February 1974, the offices of Unione Petrolifera, the industry trade association, were raided and documents unearthed which suggested that it was acting as a conduit for payments to be made by oil companies to a wide range of political parties and individuals, and that it was compiling false information on oil supplies with the intention of winning oil price increases. Foreign oil companies appeared to be heavily involved, and public disquiet was further increased by the fact that these investigations coincided with separate allegations that the CIA was also funding Italian political figures. Scandals are nothing new in Italian politics, but it is significant that these allegations shook the political world in a way that similar such affairs in the past had failed to do and the reputation of the oil industry in Italy was blackened far in excess of that in any other country except perhaps the United States. During 1973 and 1974 two of the majors, BP and Shell, decided to withdraw from the Italian market, thus strengthening ENI's hold. Both companies withdrew complaining about the difficulties of making money in a market in which the government allowed the companies inadequate profit margins. Non-Italian members of the oil industry sometimes suggest that ENI is not unhappy about seeing the foreign companies squeezed out in this way, but Italians may well find such divestments a sign that the international companies are not fully committed to Italy, and that any Italian energy policy should be built round ENI instead.

Belgium was also racked in its turn by oil-linked controversies in the immediate aftermath of the embargo. The first of these was an extreme confrontation between a group of importing companies and the government pricing authorities. Faced with losses caused by pegged product prices, practically the entire industry went on an imports strike in March 1974 to get price controls eased, including majors and independents, companies with and without refineries in Belgium, the French CFP and Elf-ERAP as well as the Belgian Petrofina. In the face of this combined weight, the government relented and permitted prices to move upwards despite opposition by unions and co-operatives. As a form of compromise, the government created a control commission with sweeping powers to investigate all aspects of oil-company activities with the intention of allaying some of the worst left-wing worries about the handling of the prices issue (*Petroleum Economist*, April 1974, p. 151, May 1974, p. 192). This did not satisfy the socialists who wanted an independent state oil refinery. The idea for the project developed in the 1960s and, by 1970, had come to include the National Iranian Oil Company and CFP. The latter dropped out but the socialists remained keen that this project should go ahead. In early 1973 there was sharp controversy when it was learned that the former Minister of Economic Affairs, Henri Simonet, just before moving on to an appointment as EEC Commissioner for Energy, had not only formally established the Iranian Belgium Refining and Marketing Company in accord with NIOC, but had also made some politically controversial appointments to the board of directors. The Belgian government coalition survived the resultant row, but dithered over the extent to which the project was economically viable and allowed a key deadline in January 1974 to pass without a decision, thus permitting the Iranians to pull out of the deal. In the aftermath of this failure, the Belgian government resigned and elections had to be held (*Petroleum Economist*, March 1974, pp. 101–3).

No one clear picture emerges from these episodes. Some, such as the scandal of political payments in Italy or the German Cartel Office investigation, clearly aroused public distrust of the foreign companies. Others, like the Schvartz Report or the Belgian refinery involved indigenous companies as much as, if not more than, the majors. All came to the surface in the immediate aftermath of the embargo. Quite simply, the events of the autumn of 1973 acted as a catalyst in transforming the world's perceptions of the oil industry, confirming that the power-shift from the industrialised world was irreversible in the short to medium term. They also created uncertainty in importing governments, acutely aware of their lack of knowledge of how the industry was being run, and suspicion among the ultimate consumers that corporate interests did not coincide with those of the private citizen. This was even found in Japan where there was a dispute over prices which led the majors to discreetly threaten to cut back supplies and an antitrust case which developed from an investigation by the Fair Trade Commission of the activities of the Petroleum Association and various companies and culminated, in May 1974, with the indictment of twelve leading oil companies, including Shell's subsidiary (Tsurumi, 1975, p. 122). Here again, there was the same kind of mixture of official unease about company pricing policies and a concern about antitrust issues found in Europe and in the United States. However, in Japan and France, at least, there were central planning bodies strong enough to shrug off attacks from antitrust authorities and maintain fairly tight central control over the activities of oil companies.

It would be misleading to suggest that the oil companies came under pressure in the industrialised world solely because of the embargo's effect. For some time there has been a trend for governments of the industrialised world to become more involved in the management of their economies and it has become increasingly less acceptable to give private, especially foreign, enterprise free run of such a crucial part of the economy as the oil industry. It was not just OPEC nations which were redefining their policies toward oil production. Canada, for instance, was increasingly dissatisfied with the US assumption that all American oil deficits would be met unquestioningly by its northern neighbour and rebelled against an oil policy structured around the needs of the international oil industry, rather than Canadian requirements.

In Europe, the discovery of oil and gas in the North Sea forced parent governments like Britain and the Netherlands to rethink their attitudes toward oil companies and brought into play a new political force in Norway. In 1976, Britain, one of the staunchest supporters of free enterprise within the international oil industry, created the British National Oil Corporation (BNOC) as a state oil company in preference to the semi-public BP. The argument most frequently given against nationalising BP was that a state oil company would have difficulties with the USA in its activities in Alaska – but it is also significant that the mixed public-private form of company ran into disfavour in other countries. France created Elf-ERAP as a stimulus to the mixed CFP, and Norway created Statoil instead of relying on the mixed Norsk Hydro. The British government also insisted that future exploration should be done with state participation and that companies with existing licences would be discriminated against if they did not accept such participation. Each field within the North Sea is now a separately taxed entity, thus stopping the practice of offsetting losses made in one field or elsewhere against profits made from successful fields. There is a new petroleum revenue tax which is a form of 'super' profit tax aimed at keeping companies' post-tax returns on investment down to the 25–40 per cent region. In addition, the British government has nationalised the sites for building production platforms and insisted that, other things being equal, oil companies should use British suppliers. It can control the routing of, and access to, pipelines and, within limits, regulate the rate at which a field is depleted.

The Norwegians have a similar array of controls and are more restrictive towards non-Norwegian companies wanting to explore for oil within Norwegian waters. Norway is worried about the inflationary impact of over-rapid offshore development, and has a strategic problem with the Soviet Union in her northern waters (Aamo, 1975). The French only allow foreign companies into offshore waters if French companies are given equal access to energy supplies in these companies' respective home territories. The Danes have had an unfortunate experience through granting all their offshore territory to one consortium whom sections of public opinion accused of refusing to produce potentially commercial gas discoveries, and are thus making sure that they insist on much tougher terms when they hand out exploration licences off Greenland. Even the Dutch, who have taken a relatively relaxed attitude towards the Shell/Esso team, have been increasing their tax rates on offshore oil and gas operations to 70 per cent in the case of gas, have extended state participation into oil exploration as well and have put an 85 per cent tax on windfall profits from Groningen gas (*Petroleum Intelligence Weekly*, 5 April 1976, p. 11).

Even where domestic production is not a major consideration, industrialised governments have strengthened the role of national companies. The Germans became unsure of the wisdom of relying too much on foreign companies during the Suez crisis and were offended by a series of bids for some German oil companies by Texaco, Gulf and even CFP. In 1969 the Federal government started offering subsidies to Deminex, a joint oil exploration and development enterprise, which eight German companies had founded to find oil abroad (Mendershausen, 1976, p. 25). A further governmental grant of DM800 million was made available for approved projects in 1974 when the company was more tightly structured with fewer shareholders and stronger state supervision. At the same time Germany has been seeking to merge the leading downstream operators, Veba-Chemie and Gelsenberg, into a company with interests in oil, chemicals, power-generation and, through their majority holding in Deminex, oil exploration. In Mendershausen's words:

> By the beginning of the 1970s, it seemed to have become a government objective that (1) the share of German-owned companies in the domestic market for petroleum products should be at least 25 per cent; (2) in addition to the small domestic crude oil base there should be a foreign crude base of some size under the management of German companies; and (3) some of these companies should be consolidated into a firm with international weight. (1976, p. 26)

By 1976, Germany was well on its way to achieving these goals.

The Japanese have followed similar policies. The Petroleum Industry Law of 1962 was an attempt to strengthen control over the local industry, but it was not until 1966 that an attempt was made to increase the supplies of oil under Japanese control. In 1967, the Japanese Petroleum Development Corporation was created as a quasi-governmental body to co-ordinate and promote oil development by Japanese companies. This is primarily a financing body which investigates possible ventures, assists in drawing up agreements between Japanese companies and then, as operations get going, withdraws. It has enabled Japanese companies to become involved in Abu Dhabi, Indonesia, Alaska, Canada, Nigeria and Zaïre (FEA, 1975, pp. 92–3). As in Germany, the government is trying to restructure the smaller domestic companies by offering low interest rates to finance marketing facilities and associated refineries, and Kyodo Sekiyu, a group of five small companies, now controls about 10 per cent of the market.

The trend is towards the creation of new state companies or the consolidation of private national companies – except in the United States where the emphasis has been on increasing competition and the idea of a national oil corporation has been given relatively little attention. There are signs that the USA may take a more interventionist role. In 1975, when it was decided to set up a national stockpile, the oil industry was bypassed, although given an option to provide around a third of the eventual stocks. It was the government which sought to make direct agreements for supplies with government producers and to administer the stockpiles which would be under government control, as would also be the preparation of storage facilities (*Petroleum Intelligence Weekly*, 15 March 1976, p. 9, 29 March 1976, p. 7, 3 May 1976, p. 5).

It is interesting to speculate on the impact of Canadian developments on US policies. Like the Germans and Japanese, the Canadians slowly came to the conclusion that they would no longer passively accept the US oil companies as arbiters of their oil policies. The switch in Canadian policy came relatively late, despite an extensive long-term debate over the impact of US corporate dominance which was particularly noticeable in the oil sector (Levitt, 1970, p. 61). By the early 1970s foreign, mostly American, companies owned 83 per cent (by book value of assets) of Canada's petroleum refining capacity, 77 per cent of the entire petroleum industry and 40 per cent of domestic Canadian pipelines (Debanné, 1974, p. 134). The controversial national oil policy of 1961 was a compromise which allowed high-cost production from the western provinces to find a US market while the eastern provinces were left to rely on cheap imports from the majors' Venezuelan offshoots. When US policy makers realised that their estimates of US domestic production had gone awry, they assumed that Canadian exports would fill the gap. There was a strong reaction in Canada and renewed feeling that Canadian oil and gas should be conserved for Canadian needs, with only the surplus being sold south of the border. Export controls on oil were imposed in March 1973 and gasoline exports restricted in the summer of 1973 when there was a real gasoline famine in the USA.

In December 1973, Prime Minister Trudeau announced his proposals for a complete change in Canadian energy policy. He decided to scrap the national oil policy in favour of a new policy in which western oil would be joined to eastern markets by extending the Interprovincial Pipeline east from Sarnia into Montreal. He also called for the creation of 'a publicly owned Canadian petroleum company principally to expedite exploration and development'. Petro-Canada was created in 1975, with government backing. Its functions were initially limited to exploration, development and the handling of federal stakes in Panarctic Oils and Syncrude. By 1976 Petro-Can was moving into exploration off the eastern coast and encouraged by the government to go into frontier resource development where it was given the right to acquire up to 25 per cent of frontier properties without paying a share of previous investment if no discovery was made (Debanné, 1974, pp. 137–45; McKie, 1975, p. 80; *Petroleum Economist*, August 1975, p. 315; *Petroleum Intelligence Weekly*, 5 April 1976, p. 9, 3 May 1976, p. 9, 24 May 1976, p. 10).

Canadian policies toward the companies would seem to be typical of the kind of discreet economic nationalism widespread within the industrialised world for the late 1970s and early 1980s. Although there has been no serious attempt to oust the majors from their traditional activities, there seems to be a distinctly expressed need for a state presence in newer developments – be they offshore exploration in the North Sea, frontier development found in northern Canada or the setting up of a national stockpile in the USA. But even in these newer areas, governments generally offer the majors some participation. Governments appear unwilling to stir up controversy by nationalising operations in which the majors have been traditionally involved but find scope for bargaining with the companies in newer activities. The Canadians are aware of the role that pricing policies must play if both public and private industry are to be self-financing in areas requiring substantial future investments and have suggested that the government may need to play a more active role in monitoring the cash-flow of private companies to ensure that an appropriate share of funds goes into exploration and development.

An era is ending for the majors in both the host and consuming nations. Their freedom to do business is being eroded by a general increase in political and economic nationalism, and expansion by acquisition is getting difficult because they have grown too large and too visible. Consumer governments will probably attempt to reserve 30–35 per cent of the domestic market for indigenous companies, private and state owned. The majors' share of new energy fields will be constrained by demands from governments for national participation, probably as majority shareholders, from the start. They will increasingly find themselves hemmed in by price controls and by closer government monitoring of transfer pricing, investment policies and technological skills. However, there is no sign that increased government involvement will drive the traditional companies out of business. It seems reasonable to assume that most governments will set price levels which will allow the companies to generate the profits needed to guarantee investment in new sources of energy.

Governments realise that the activities of the companies are too politically sensitive to leave entirely without regulation. On the other hand, the growing sophistication of the debate on energy policy at both the national and international levels is such that there is a growing realisation of the interconnection between regulatory decisions aimed at protecting today's consumer and the investment decisions which will affect the coming generation. The result is that there is plenty of scope for sensitively presented arguments from the oil companies aimed at winning quite reasonable treatment from the regulatory authorities. Although the relationship between companies and industrialised governments will increasingly be dominated by the latter, in so far as the mixed economies of the West survive, the majors will survive as some of the most dominant companies within them.

NOTES

1 An informed source claims that Walter Levy wrote much of the speech.
2 According to Rustow, only about one-twelfth of Arab exports could be carried in Arab ships when current orders are completed.
3 Tax Reduction Act of 1975 (*Petroleum Economist,* May 1975, p. 174).

10
Will the Companies Survive?

Despite considerable government encroachment, the historically autonomous, traditionally privately owned oil companies will survive. The environment in which they will be working will, admittedly, be very much more complex than in the past, with competition from a wide range of new entrants into the industry, both state owned and private, but the majors will draw on their long-standing relationships with various countries and their managerial strengths to ensure that they maintain a dominant position. At the same time, though, they will have to work within a new framework in which plant location and the distribution of markets will increasingly depend on intergovernmental agreements (in the past, such issues were primarily determined by the interplay of the majors in a relatively oligopolistic market).

CHALLENGE OF THE NATIONAL OIL COMPANIES

The most obvious challenge to the oil industry is going to come from the array of national oil companies which have been the prime beneficiaries of the growing host participation in the activities of the former concessionaires. The events of 1973–4 meant that Petromin (Saudi Arabia), NIOC (Iran) and KNPC (Kuwait) became three of the seven largest oil producers in the world, while a number of other national oil companies also increased in importance (Jacoby, 1974, p. 190). Some, like Mexico's Pemex, had been running for decades, others like Petroven (Venezuela), Sonatrach (Algeria) and Pertamina (Indonesia) were younger, but were still forces with which to be reckoned. These companies have been assigned many of the assets previously controlled by former concessionaires. Increasingly they control domestic marketing, most local refining and oil production and are moving into the international markets. The key question is the extent to which they can build on the potential national monopoly of oil production to move into the downstream operations in which the private companies are now concentrated. This depends very much on the degree to which the older established companies can identify managerial and technological strengths which the newer state companies will find difficult to duplicate. If they can convince producer governments that this is the case, then the challenge of national oil companies will be less troublesome than would appear at first glance.

TECHNOLOGICAL LEVERAGE

Nationalising the production of crude oil is a relatively easy step in the producing governments' campaign to oust the former oil majors from their position of dominance. It is far more difficult to develop state companies and agencies which can expand the range of their activities beyond oil production without dissipating the financial returns which the producing governments have boosted over the last five or six years. Although many Western commentators underestimate the management skills of the developing world (a notable demonstration of this tendency was in 1956 when the Egyptians nationalised the Suez Canal, and it was widely held in the West that they would be unable to manage it), many Third World commentators gloss over the importance of management skills and the difficulties of acquiring them. This is particularly true in the oil industry where host country 'take' will generally be some ten to twenty times the actual cost of producing oil. Unless there is very close control, there is scope for quite spectacular inefficiency. The story of Pertamina is an object lesson of how a national oil company can get out of control with almost disastrous results.[1] Indonesia's state oil company was run as the personal fiefdom of one man, General Ibnu Sutowo, who went on a gargantuan spending spree as the oil money flowed in. The company went into real estate (worth some $500 million), an airline (a hundred planes and sixty helicopters), oil tankers (at least twenty-eight in the 150,000 ton range at the time the bottom dropped out of the tanker market), a rice estate, a steel complex and a fertiliser plant. It was planning to build a harbour and refinery as well as to develop a telecommunications arm. Despite the fact that Pertamina earned 49 per cent of Indonesia's foreign revenue in 1974, it went massively into debt. Repayment difficulties brought publicity and, by the time all the various deals were uncovered, it looked as though it had run up debts to the tune of $10 billion (*Business Week*, May 1976, p. 63), an enormous figure for a commercial operation. It is over twenty times the loss made in 1975 by Singer, the biggest money loser for that year among *Fortune*'s list of the 500 largest industrial companies, while Exxon, which was the most profitable company during 1975, made a mere $2·5 billion profit in comparison (May 1976, pp. 318*ff*.).

National oil companies are bound to be very large in comparison with other indigenous companies and agencies. One lesson of the Pertamina disaster is that there should be extremely tight legislative constraints on the areas into which state oil companies can diversify. It is not much use to have a company which can produce oil efficiently, if it promply loses most of the profits by, say, investing in the tanker market at the wrong time or building refineries without ensuring markets for their products. The central question is in what areas the new companies can hope to replace the former concessionaries without much loss of efficiency, and in what areas they will find it difficult to compete. It is important to remember that the oil industry consists of a variety of operations making a widely different set of demands. The technology in each area matures over time so that operations which could originally only be carried out by a handful of companies come to be common knowledge. The national oil companies will find it easiest to enter the more mature technologies, while the traditional majors will seek to dominate those fields which are the fastest moving and the most technologically complex.[2]

At one end of the spectrum are mature technology activities such as crude oil transportation, much land-based crude exploration and production and oil refining. In most of these it is relatively easy to buy the necessary equipment and hire the experts. Even before the current slump in the tanker market, which has made entry extremely easy, anyone with money could acquire and run the largest of crude oil carriers. If an oil refinery was wanted, it was relatively simple to approach established process plant constructors like Kelloggs, Bechtel and Foster Wheeler and purchase a perfectly adequate turnkey oil refinery (the constructors build it, handing over to the customer a plant which can be immediately put on stream). There are few insurmountable problems in nationalising oil industries. Kuwait and Saudi Arabia control oilfields whose workings are well understood by anyone with adequate training and experience in reservoir management. If these governments, or others like them, want to build a refining or basic chemicals industry, there should be no technological reason why they should not do so.

Where the companies come into their own is in capitalising on their experiences at the frontiers of technology. It is very much the majors who dominate oil exploration and production in the inhospitable regions of Alaska and northern Canada. They are equally strong in offshore exploration. About a fifth of the exploration licences now held by Exxon are in areas with water depths in excess of 600 feet (the commercially exploitable fields in major offshore oil provinces like the North Sea are all currently at lesser depths). Companies like BP, Exxon and Shell are developing an expertise in the offshore area which smaller, newer companies cannot match, so that any country wanting to exploit hydrocarbons off its coast naturally turns to the major oil companies. The British government has had to deal with them, despite its desire to build up the strength of the British National Oil Company. Even the USSR, with a large, well-established industry, still finds it necessary to negotiate with companies like BP for assistance in deep-water drilling in the Caspian Sea.

Offshore exploration is an extreme case of the majors maintaining their technological edge, but there are other areas as well in which their expertise pays off. For instance, as individual fields grow old, the need increases for enhanced recovery techniques such as the injection of inert gases into the field to help force the oil out, or the addition of chemicals to render the oil more viscous, thus increasing the efficiency of injected water. These secondary and tertiary recovery techniques are very much the province of American companies, largely because the USA has many nearly depleted fields where such techniques are profitable. When OPEC countries, such as Iran, are in the market for such techniques, they may be forced to turn to companies like Exxon. The liquefaction of petroleum gas is a new technology which is still substantially dominated by international companies like Exxon and Shell, though the technology is in the process of passing into the realm of general knowledge. Then there is a company like BP which has pioneered the process of producing proteins from oil and with which it is necessary to come to an agreement if, as in Venezuela, a government decides to enter this field.

It is by continually identifying and dominating such frontier technologies that the companies will survive. Despite the US role in the Vietnam War, the North Vietnamese have told US congressmen that they particularly want American companies to help in developing offshore oil resources (*Petroleum Intelligence Weekly*, 5 January 1976, p. 8). Norway, which has been wary about allowing the majors into its northern waters, now appears convinced that Statoil must co-operate with experienced oil companies to ensure strict environmental safeguards and to gain access to the latest technology (*Petroleum Intelligence Weekly*, 10 May 1976, p. 1). Brazil, where oil activities have long been Petrobras' private preserve, has invited international companies to bid for exploration and development rights (*Petroleum Intelligence Weekly*, 28 April 1976, p. 7). Libya, having nationalised some of its oil production, is extremely reluctant to do the same to the much more complex liquefied natural gas plants of companies like Exxon, which require both sophisticated technology and access to international markets (Ghadar, 1975, pp. 3–13) – a striking case of how command of advanced technology can protect oil companies in even the most hostile environment.

The implications are that the traditional companies will be able to use their technological strengths as bargaining cards with both producer and consumer governments, but that the exact balance struck will depend on the specific needs involved. Kuwait has but a limited chance of finding new reserves in its small territory and needs exploration skills less than Iran, which, in turn, needs more secondary and tertiary recovery techniques than Saudi Arabia, which, in turn, needs more expatriate help with its industrialisation programmes than Venezuela, which may well at some stage call in some outside help in developing its large deposits of heavy oil – its *Faja Bituminosa.* These countries will not automatically call on the majors for help, since there will be a growing number of sources on which to draw, but the majors can expect to provide a reasonable share of such assistance for reasons of historical inertia and because they are well placed to draw on an extremely diverse range of experiences from all parts of the globe.

The traditional companies have a further advantage in that they still dominate the refining industry of West Europe and North America and hence control the largest markets. In the case of Europe, the seven largest companies still controlled 55 per cent of refining capacity in 1972 (down from 65 per cent in 1953) and, with significant overcapacity forecast to last into the early 1980s, it would seem unlikely that their hold will slacken very much further in the immediate future (Jacoby, 1974, p. 194; *Petroleum Intelligence Weekly*, 26 April 1976, p. 6). The national oil companies can break into the refining area in three ways. First, they can increasingly supply the refiners of the rest of the Third World. Secondly, they can go in for refining themselves, seeking to go over the head of the majors by selling directly to ultimate customers, be they electrical utilities or independent petrol station chains. Thirdly, they can insist on selling crude oil only to companies which are willing to take a specified proportion of refined products. There is not a great deal that the majors can do about holding on to Third World markets, but the prospect of national refineries supplying world markets is something which worries them. Refining projects planned in the Middle East and Africa are due to contribute some 27 per cent of the planned 1980 increase in world refining capacity, despite the fact that this industry will probably only be working at 76 per cent capacity by then (*Petroleum Intelligence Weekly*, 26 April 1976, p. 6). On strictly economic grounds national oil companies may find it difficult to sell more than a portion of the refined products their plants will be capable of producing. The cost of building refineries in the Middle East is quite high and transportation costs will remain high until giant product tankers are accepted on world waterways.[3] The economics of a refinery depend on its being able to dispose of all its products which are designed to fit the needs of specific markets (i.e. in North America, refineries produce a higher proportion of gasoline than in Europe). There are doubts that refineries built so far from ultimate markets will be able to guarantee the regular disposal of the full range of their products, so crucial to their profitability. Admittedly, modern communications make it easier to locate buyers for oil products but it is still generally more convenient and flexible for an independent buyer in Europe to use the Rotterdam or Italian refining centres for marginal supplies than to rely on a centre some 6,000 miles or so distant.

It may be that national oil companies and Third World governments will decide against building refineries which could only be profitable if their feed-stocks are priced below world levels. On the other hand, internal political pressures to build refineries will be quite strong, as OPEC members believe that they have been kept in a dependent and inferior position by the old international economic order; their own refineries and petrochemical plants are symbols of independence.

For political reasons, there may be a glut of refineries and producer governments would then be forced to pressurise the traditional companies to take more of their products from these refineries. How quickly this might happen is another matter. On the one hand, being forced to take unwanted oil products will encourage traditional companies to favour crude oils from sources which do not make such conditions. On the other hand, if politically inspired overinvestment in refineries renders the whole sector uneconomic, then the traditional companies may be tempted to cut back on refinery expansion in the industrialised world in favour of allowing national oil producers to make the uneconomic investments. Here, though, the industrialised governments might step in, for there is no point in aiming for self-sufficiency in energy production merely to replace this with a growing dependence on imported crude-oil products.

The likelihood, then, is that the traditional companies will gradually get out of marketing simpler crude-oil products for the simple reason that the investment climate will become too political. However, in the short to medium term, they will still be formidable potential customers for crude oil since they alone have the refining capacity of a million barrels a day and upwards. The combination of this massive purchasing power and their technological lead should ensure that, for the medium-term future, they will continue to get preferential access to crude oil from most producing countries in the form of service and marketing contracts.

To become independent, the national oil companies will have to 'go multinational'. One strategy, more talked of than implemented, would be to buy their way downstream. This could involve the creation of shipping companies like the Arab Maritime Transport Co., or joint ventures like that between Iran's NIOC and BP (the Irano-British Shipping Company). It might also involve participation in refinery projects closer to ultimate markets, partly to gain downstream experience. (NIOC was tempted by one such deal in Belgium.) It is clear that part of the thinking behind the putative deal whereby the government of Iran was to buy 10 per cent of Occidental was that this would lead to a series of joint ventures which would give the Iranians further experience in the downstream area – and participation in Occidental's under-utilised refineries would have been one possible target.

There are drawbacks to this approach. The first is a developmental one. The national oil companies may get out of control, becoming too interested in the prestige of overseas investments which may well be economically questionable but attract all the most ambitious national managers and planners who should be concentrating on the wider developmental needs of the producing country (i.e. investing in a tanker fleet may be socially undesirable compared to investing the same amount of capital and skilled manpower in reinvigorating the agricultural sector of the country).

A second drawback is the assumption that national oil companies should base themselves on the integrated model of the majors, and because the majors have controlled all stages of the oil process from exploration to marketing, they should go into tankers, refineries and product marketing themselves. Times have changed. Integration made sense in the earlier decades of this century when the international business environment was still somewhat underdeveloped and companies discovering Middle Eastern oil had to go into tanking themselves because there were no independent shipping magnates like Onassis to do the investing needed in such a volatile industry. There were few independent refiners until the late 1930s. The majors had to invest in petrol station chains and other marketing operations because oil was only gradually finding wider acceptance in the industrialised economies. Coal was still dominant as a source of energy in much of the industrialised world as late as the 1950s, so companies with large concessions in the Middle East had to market aggressively in order to get the volumes necessary to give them profits – and keep producing governments happy.

Today there are an increasing number of entrepreneurs willing and able to step in to carry out many of the functions which were once virtually the sole prerogative of the majors. In shipping, there is the generation of shipping tycoons; in refining, a growing number of independent companies willing to risk building refineries in centres like Sicily or the Caribbean; in end-marketing, a far wider number of relatively small businessmen who have learned to run a limited number of very high throughput petrol stations at a profit. The official philosophy of OPEC is that oil should not be produced at the maximum rate, but should be conserved, with the returns to the producing government coming from improved 'take' per barrel of crude oil, not from expanded output. The two main planks underlying the integration of the majors are missing in today's circumstances, and there are good grounds for suspecting that a company which concentrated on producing and marketing crude oil would give a better return than one that followed the tankers-to-refinery route, despite the current cheapness of tankers. It is worth noting the ease with which losses can be made in downstream activities. Burmah Oil went virtually bankrupt at the end of 1974 due to ambitious attempts to buy into the American market and to substantial losses from overcommitment in the tanker field. An ill-timed and badly located refinery in Newfoundland (Shaheen's 'Come-By-Chance plant') went bust in 1976 after taking a lot of subsidies from Canadian sources, dragging into trouble an overambitious Japanese supply agent which had moved in after BP had dissociated inself from the project. West Germany's state-controlled Veba lost about $175 million on its oil operations in 1975, with losses of around $1.47 per barrel sold. The Belgian company Petrofina claimed that nearly all product marketing in Western Europe was unprofitable in 1975 and that, through restrictive price controls, it lost $25 million in both 1974 and 1975 within Belgium alone (*Petroleum Intelligence Weekly*, 8 March 1976, p. 9, 29 March 1976, p. 9, 31 May 1976, p. 6).

The traditional majors often feel that integration gives them few advantages – and plenty of hostages to fortune. Providing they can win reasonably priced bulk contracts for crude oil, they can operate their refining operations just as effectively as if they were still producing crude oil themselves under their concessions. They can still purchase crude oil from a wide enough variety of sources to give them the flexibility needed to programme a logistical system which, in the case of Exxon, has involved manoeuvring 500 ships from 115 loading ports to 270 destinations, carrying 160 different crude oils between 65 countries (Sampson, 1975, p. 8). It is the smaller refiners who run into difficulties should crude oils from a particular part of the world or of a particular chemical characteristic be lost; giants like Exxon can much more easily adjust schedules so that alternative crude oils with similar characteristics are fed into the right refineries. On the other hand, involvement in extractive industries is still perceived as controversial and somehow 'illegitimate' (Boddewynn and Kapoor, 1972, pp. 436, 447) so that even if they do capitalise on market strengths and technological expertise, they are still vulnerable to potential discrimination.

The tempo of technological diffusion seems to have been speeding up within the industry, and companies are increasingly aware that a breakthrough they may make, say, in refining technology, will be replicated by other companies elsewhere in the world before they have maximised their return from it. Moreover, the international complexities of the industry make it extremely difficult to plan for some sectors. For instance, in the shipping field, how many other oil producers are going to go along with Nigeria in demanding the right to provide tanker transport for 50 per cent of crude oil purchases? With the build-up of the OAPEC joint Arab tanker company, and with Iran co-operating with BP in the shipping field, the practice of insisting on some ratio of carriage by producers is bound to spread. Will the industrialised world then insist on a certain proportion of imports being carried in vessels of 'secure' nationalities, as the USA now insists that coastal trade be carried in American flagships (which can add to the costs of transport)? (*Petroleum Intelligence Weekly*, 15 March 1976, p. 5, 17 May 1976, p. 10.) When a glut of refining or petrochemical capacity arises, will this not call for some form of inter-governmental deal to forestall a trade war? Given the tendency for producer governments to insist increasingly on companies committing themselves to minimum levels of crude oil offtake, what happens in the next industrial down-turn when oil will be flowing from both Alaska and the North Sea? Might not the companies be under conflicting instructions from, say, the British, American and OPEC governments?

The oil industry will undoubtedly be one of the battlefields in the struggle of governments at both ends of the development scale to influence the workings of the International Division of Labour. Classical western economists have been under challenge for some time, but their assumption that the Invisible Hand will dictate a rational location of activities within an international industry like oil seems hopelessly old-fashioned. The 'hand' has become distinctly incarnate and, on occasion, definitely mailed. The major industry so far where intergovernmental agreements have formally tried to slow down the migration of an industry has been textiles, where the Multi-Fibre Arrangement has to some extent protected the industrialised world from the too-rapid decimation of an industry which has been undercut by lower costs in the underdeveloped world. The oil industry is different. The creation of refining industries in oil-producing economies is not happening because production factors overwhelmingly favour this location (as textiles naturally gravitate toward cheap-labour economies), but because oil-producing governments have won themselves the political and financial clout to insist on a redistribution of the industry's centre of gravity whatever the strict economics of the case. The strategic implications of this shift assure that the tensions will arise not only in the industry but in the international political scene. The oil companies would obviously be well advised to reduce their visibility, particularly in those parts of the world where governments are not going to fight on their behalf. The evidence is that they are working on this assumption. In the survey by Chase Manhattan of the investment pattern of the 'free world' petroleum industry in 1975 (Chase Manhattan, 1976, pp. 14–19) 69 per cent of the investment in exploration and production was made in North America and Western Europe. In particular, spending within Western Europe was 44 per cent higher than the combined total for the Middle East and Africa, jumping 51 per cent during 1975 to $3·6 billion. A decade previously, Europe accounted for 3·2 per cent of global exploration and production; by 1975, this had risen to 19·6 per cent. The survey also shows the way that companies are moving into the chemical industry, spending, over the past decade, somewhere between 20 and 25 per cent of their investment in this area.

What is happening (these investment figures illustrate a trend which started over a decade ago) is that the traditional companies have given up the role of integrated companies and have instead entered on a massive search for those areas in which they have a strong competitive edge on rivals. More and more, they are in the business of identifying their operational strengths in one part of the world and 'multinationalising' this as quickly and as thoroughly as they can to other parts. If they conclude that there are no worthwhile profits to be made in certain sections of the oil industry, or even in the whole of it, it is more than likely that they will evolve into new kinds of companies bearing only a historical link with the majors the world knew in the 1930s and 1940s.

One of the first signs that thinking had changed within the industry was when some companies started to divest themselves of activities which they judged would not become profitable fast enough. BP and Shell both reacted to Italian price controls and sold off their Italian subsidiaries over 1973–4. Exxon pulled out of India and the Philippines. During 1973, Gulf largely withdrew from the German market, selling off a refinery and over 700 service stations to Veba. These divestments – unusual for an industry used to steady growth – should also be seen in the light of the considerable amount of organisational devolution which had been going on within some companies. Gulf regrouped its worldwide operations on a functional basis in mid-1975. In place of the former geographical divisions, six 'control' companies were created to act as investment centres with wide powers and responsibility for their own financial results, covering the extraction of oil, gas and other energy sources, oil refining and product marketing, transportation and sales to third parties, chemicals, technology and research and real estate. The implication was that the extractive industries company, Gulf Energy and Minerals, would be free to buy and sell competitively, while the refining and marketing company could buy oil from outside the group if it could get better terms.[4] This type of evolution of an oil company's structure is a natural response to the growing political and commercial pressures. Both Conoco and Sun Oil followed Gulf's example of decentralisation, though most companies have satisfied themselves by imposing considerably tightened financial control on all aspects of their operations (*Petroleum Economist*, October 1975, pp. 379–81).

Greater concern about profitability is shown not only by a readiness to pull out of unprofitable situations, but also in the search for future areas of growth in which they can capitalise on their strong points. Their experience with oil has given them expertise within the general energy market, as well as in exploration and production techniques involving mineral sources. It is thus only to be expected that they should move into the coal industry, which calls upon their expertise in both these areas – as does the search for uranium. This widening of their horizons in the minerals field has involved them in ore mining and metallurgy in general, and their interest in the energy field has brought some of the larger companies into nuclear engineering (with pretty disastrous results as far as Gulf and Shell are concerned).

To be successful, diversification should build on real strengths. Through their refining activities they have learned about the chemical properties of oil and have thus invested heavily in the chemical sectors, gradually moving up the scale from basic petro-chemicals worth 17–20 cents a pound to the more sophisticated varieties which are worth some 20–30 cents a pound (basic crude oil is worth only 3–4 cents in comparison). Occidental, Shell and Exxon are particularly far advanced in petrochemicals with, respectively, 30, 9 and 6 per cent of their sales in 1975 coming from this sector. Shell's turnover in chemicals was worth just under $3 billion – a figure bettered by only twelve traditional chemical producers (Stobaugh, 1976).

The offshore area is likely to contribute heavily to the companies' future profitable survival. This is *par excellence* an area calling for frontier technologies in which a company's expertise can be measured by the depths at which it can carry out various operations. Exxon holds the record for operating a production platform in 850 feet of water in the Santa Barbara channel off California, and will probably hold it until 1978, when Shell plans to install one off the US Gulf in 1,040 feet of water (*Petroleum Intelligence Weekly*,

5 July 1976, p. 6). The technology involved is developing fast. In the early days of exploration in the Gulf of Mexico and Lake Maracaibo, only slightly modified land-rigs were used in shallow water, and for a long time the equipments used, even in deep water, were just a series of heavily reinforced versions of land-based designs. There are still in use giant steel offshore production platforms which are direct descendants of the early ones used in the 1940s, but there are also established concrete designs, and there will soon be a growing demand for tension-leg platform systems, in which the platform floats, tethered to the seabed by high-tension wire, which will permit production to take place when conventional platforms, resting on the ocean floor become too expensive (the latter's costs rise exponentially with increased depth). In the background, there is the further development of seabed production modules, which should make it possible to transfer a growing proportion of the production process from the ocean's surface to its bed, thus greatly reducing the need for substantial structures linking the two spheres of operations.

There are other attractions to such offshore expertise. The companies should find it relatively easy to move into the discovery and exploitation of other minerals to be found offshore. Although some other companies are interested in the manganese nodules found on large areas of the sea floor, no other industry has been forced by commercial or political necessity to develop its offshore capability as early as or on a scale comparable with the oil industry. The development of offshore technology is opening up the oceans, and the companies which are pioneering this area have no strong feeling that they are 'oil' rather than 'mineral' companies. A rapid incursion of oil companies into the preserve of traditional mining companies as far as offshore developments are concerned seems natural, for the simple reason that they control the relevant technology – just as the micro-electronics companies are eating into the market share of mechanical watch manufacturers. When technology changes massively, it is the companies with the know-how which sweep the board, not the companies which have been the traditional market leaders.

Another advantage of offshore expertise is that it takes the industry closer to the commercial exploitation of areas of the seabed which are now beyond national jurisdiction although the UN Law of the Sea conference may result in some form of International Seabed Authority. It seems highly likely that a good part of the exploitation of this area will be left to commercial companies – perhaps under licence from an international authority, with some form of royalty going to the coffers of the less-developed world. It is unlikely that such an authority would follow an extremely restrictive policy on the rate of deep sea exploitation, especially if a proportion of the profits go to Third World countries which tend to back production policies generating rapid rates of return. In fact, it is in areas beyond the sovereignty of militant national authorities that a potentially stable future for the companies may be found. On land, the problem for the oil industry is that it has to work with governments which are restricting its freedom, and there is no hope of reversing this trend in the policies of Third World producing or industrialised governments. Offshore activities offer an opportunity for at least short- to medium-term freedom.

SECOND CENTURY

Before looking ahead further, it is as well to remember how stable the traditional company has shown itself. The oil industry is now roughly 120 years old, while Rockefeller's original Standard Oil from which Exxon, Mobil, Socal and others derive was created in 1870. Right from the start oil was important and, under Rockefeller's guidance, it had a major impact on the public almost from its creation. What is sometimes not obvious to outsiders is how dominant the oil industry was in the United States even in the 1920s, and how stable the balance between these various companies has actually proved. A study of the twenty-five largest US industrial corporations in 1929, measured by assets, showed that Jersey Standard was in second place, Indiana Standard in fourth, Socony in seventh, Texaco in ninth, Socal in tenth, Shell Oil in thirteenth, Gulf in fifteenth and Sinclair Oil (now part of Arco) in sixteenth position. Apart from the fact that they slightly improved their positions relative to non-oil companies (i.e. Exxon/Jersey Standard became first in assets), the 1976 picture is impressively similar. Arco has taken the place of Sinclair, and Conoco, Phillips and Tenneco have entered the top twenty-five. Some of the non-oil giants fared badly. The 1929 leader, US Steel, slipped to twelfth place, with other steel companies falling back as well. Anaconda, the mining company, was down to seventy-first (and is merging with Arco), while the meat-packing companies (Armour and Swift), movie-industry holding company (General Theatres Equipment) and the railroad equipment company (Pullman) all did proportionately worse. Apart from that, most of today's big names are easily identifiable (General Motors, Ford, General Electric, Du Pont, International Harvester, etc.), bearing witness to the way that such companies have been able to go through world slumps, world wars and the de-colonisation of much of the world and yet still maintain their corporate existence (*Fortune*, May 1970, p. 258).

Adaptability is thus a part of the giant corporation's stock-in-trade. After all, Rockefeller was originally in the business of providing kerosene for lighting in the period before this was done by electricity, decades before Henry Ford created the mass automobile industry which would demand plentiful supplies of gasoline. The fact that electricity was developed did not mean that Standard Oil went out of business. It adapted, found new markets and went from strength to strength, until the Sherman Antitrust Law led to its being chopped into pieces. Even then, the newly created progeny proved every bit as expansionist and adaptable as the original company.

Obviously, there is no guarantee that the oil companies will escape the relative decline of their railroad and steel counterparts, but it does seem fair to argue that today's managers are now very much more aware of the dangers of restricting a company to an ageing, traditional business activity. The railroads declined because they did not (could not) diversify into newer, faster growing forms of transportation such as flying. The oil industry is fully prepared to move into businesses such as petrochemicals, coal production or nuclear engineering if that is the way to corporate survival.

Looking back over the past century, it appears that it would take a major change in social organisation before the oil companies were disbanded as corporate entities. If anything, the odds are that in the mid-twenty-first century, the major oil companies will still be high amongst the top twenty-five industrial organisations of the world. There may well be newcomers from the ranks of the national oil companies but this will depend on their ability to cope with the inevitable day when their national oil reserves become depleted. Although it is an impressive fact that NIOC's turnover was the third largest for an industrial company outside the United States in 1974, it is still of minimal importance in any activity other than the production of oil in Iran. It will not be until we see if NIOC can become a significant force in world chemicals or offshore technology that we will be justified in believing that it will be up there with the Shells and the Exxons in the year 2059 (the bicentenary of 'Colonel' Drake's discovery in Titusville).

The national oil companies have two main strategies for which they can opt. First, they can ape the majors by building up their expertise in all stages of the oil industry, moving into more and more complex downstream activities such as petrochemicals. We have discussed earlier some of the economic costs to them if they follow such a policy, but it is also likely that they will be hampered by continuous hesitation on the part of parent governments, not quite sure that so much attention should be paid to developments half the globe away. Already the Norwegians are attempting to limit the freedom of their national oil company to operate abroad and it is reasonable to expect that OPEC governments will increasingly make similar reservations. Having a stake in two or three licence blocks in the North

Sea may give a company like NIOC the impression that it is becoming fully competitive in the international arena. However, until it is willing (or is given the freedom) to put in major bids for oil rights throughout the world, it is always going to come off second-best to the majors, whose very existence depends on their snapping up and exploiting opportunities wherever they may present themselves.

However, the national oil companies can opt for a second strategy, playing a marginal role in the international industry but becoming leading industrial conglomerates in their own economies. This alternative strategy looks more likely and they could end up with a monopoly over oil production, refining and distribution, much of their country's shipping industry and the chemicals industry. If the model of ENI in Italy is illuminating, they could well find themselves taking major stakes in construction, textiles and even nuclear engineering industries. Obviously, political factors will determine the scope given to each national oil company to expand. We have already seen the Indonesian case of Pertamina which expanded far too fast into non-oil areas. On the other hand, the case of Mattei in Italy shows how a forceful personality and skilled politician can turn a not particularly big oil concern into a country's dominant industrial force. A pattern has been set by Pemex, which has chosen over the last thirty years to build itself into the most important enterprise in the Mexican economy.

The future international economic environment of the oil industry runs counter to the hopes of the pioneers who created bodies like GATT. The majors are not going to regain the freedom they had as late as the 1960s to invest and trade very much as they pleased. The oil world is moving away from the free-trade, free-investment pattern which the dominant Western powers imposed on the world earlier this century, toward a two-tier economic system, in which a central, freely competitive sector is increasingly circumscribed by the actions of national governments, simultaneously seeking to influence the location of investment and the pattern of sensitive trading flows, and also to build companies or agencies at the national level which will have sufficient weight to act as countervailing forces in negotiations with the operators in the international arena. In this latter sector, a new balance is in the course of being struck between national self-interest and the workings of international comparative advantages. The exact point at which this balance will be struck will reflect the resurgent determination of national governments to put narrow self-interest higher in their order of priorities. However, this balance will also be affected by the skill with which the international companies will deploy their technological and managerial strengths.

There are various reasons why the balance may well be struck with less emphasis placed on economic efficiency and more on a variety of national quasi-economic goals. For one thing, the governments of the industrialised democracies are increasingly aware that their electorates judge them on their ability as economic managers. Since the workings of the international economy affect voters both directly (plant closures) and indirectly (the feeling of powerlessness when corporate headquarters are far away; xenophobia in face of foreign takeover bids), the governments of the industrialised world have to become involved. At the same time, the Third World now takes it as an act of faith that the international economic order is skewed in favour of the rich, and its governments thus reject *laissez-faire* principles while they try to redress the balance by political action. In the case of oil, government intervention is even more inevitable, given the extremely dominant role of the state in the countries concerned, which makes it less likely that decision makers will leave the strategic planning of such a key industry to private foreign companies. All these factors interact. Once one set of governments has become involved, the others inevitably follow, since, as OPEC has shown, a group of like-minded governments is able to influence the distribution of the industry's gains.

The interesting question for the oil industry is where increasingly activist governments will draw the line between narrow economic efficiency and wider nationalist considerations. The indications are that they are tending to move toward a situation in which they (or indigenous companies working on their behalf) will initiate or control most major investment decisions within their domestic economies. Often such policies will clearly be both politically and economically justified but there will be occasions when ill-thought-out nationalistic goals will lead to bad economic decisions – when an investment's timing will be wrong, its products unmarketable or its location unfortunate. Certain industries will attract such misguided investments more than others. Those concerned with defence always tend to be heavily subsidised in the interests of national security. Indigenous investment in high technology will be made in the name of national 'modernity'. Those connected with natural resources will be sought to guarantee that foreigners will not rape the national heritage.

In the face of such basically political investment, companies can only protect themselves by refusing to invest in such sectors themselves, since their profitability, and hence their survival, depends on their operating in sectors where the balance of investment in productive capacity bears a reasonably close relationship with ultimate demand. The politically motivated investment made in such industries will ensure that there will be a perpetual state of overinvestment, boding ill for companies without generous national authorities behind them. Governments are not as constrained by strict profitability as companies. They can consciously or unconsciously subsidise uneconomic projects; they can use general tax revenue to write off losses.

However, there will always be some point at which even the most nationalist government becomes aware that the costs of autarchy in certain projects are politically unacceptable. For one thing, with the exception of the handful of oil states with revenues well beyond the cost of satisfying their populations' needs, all governments are faced with the fact that finance and skilled manpower are in too short supply to be tied up endlessly in projects which could be finished more cheaply and efficiently by calling on outside help. Then, again, there will come a point at which delays in completing projects can become too politically embarrassing to tolerate. As all societies become more open it becomes harder to disguise the fact that similar projects are being completed faster in other countries. Strict considerations of efficiency cannot be forgotten entirely – and that is what will ensure the long-term survival of the traditional, private oil companies, as long as they successfully concentrate on dominating the most commercially viable technological frontiers. The extent to which modes of operation must change will vary from country to country, and it will often be necessary to demonstrate that they are positively contributing to national economic well-being.

In most cases the traditional company will move from being integrated to a form which is best described as a 'skills bank'. As governments become more directly involved with the overall strategy of major investments, international companies will find themselves in a supporting role, supplying markets, management, technology and finance as circumstances dictate. No longer will they be free to view certain parts of the world as 'their' territory in which they can invest as they see fit. Instead they will have to evaluate a much wider range of commercial opportunities, an increasing number of which will come in the form of calls for tenders by national governments or their agents from companies willing to provide part of a package of operations. This will force them into a semi-banking role of evaluating the worth of investment proposals put to them by potential clients. However, unlike banks, which prefer to limit their involvement to providing finance, the oil companies will want to capitalise on a much wider range of assets. They will choose investment opportunities which will give them the greatest return on the full array of their technical, marketing and managerial resources. If they can find opportunities as well for taking complete management control of a venture, they will be delighted, but their chances of finding such openings within the most basic parts of the oil industry are diminishing. In most of their operations they will have to settle for a diluted management control in the form of joint ventures or management contracts. The pattern of ownership in each set of operations will vary according to circumstances, and the degree to which the majors can co-ordinate or integrate the various operations in which they are involved will depend increasingly on the wishes of the various governmental or para-governmental partners with which they will be co-operating in various parts of the world. What is good for Exxon may not find favour with oil-producing governments with which it may be a partner – and it is these authorities which will increasingly set the terms for such joint ventures.

Companies will still try to integrate some of their activities since there are sound economic reasons for linking parts of the oil industry, but this integration will come about through a series of agreements with various partners and not through the sole decisions of a single company. A model of the type of pattern which could emerge comes from the car industry, where Fiat is involved in projects with various East European concerns (Yugoslav, Polish and Russian). Although Fiat does not own any of the plants, there is a certain amount of specialisation between them, with flows of components moving from one to another, as well as to the Fiat empire outside the Eastern bloc. In the case of oil, one might find crude oil produced from an operation with which a major has a management contract to provide production expertise assigned to a state-run refinery, then to the ships of a tanker fleet owned half by the major and half by the producing government, and ultimately to a marketing operation in which the state might have a 10 per cent interest.

The success and failure of the central management of such complex corporate arrangements will no longer rest on an unchallenged right to co-ordinate the global flows of oil, but on an ability to exploit fully the global potential of the managerial and technological expertise which is cumulatively gained over the range of projects with which the company is involved. Management will not just seek to improve immediate financial returns from each involvement, but will try to enlarge the companies' stocks of exploitable skills, having to rely on negotiation rather than decree, and needing to come to terms with a world in which many of its former subsidiaries are independent competitors as well as potential clients for various services.

The emerging pattern will be much less clear-cut than in the days of globally integrated companies. At the centre will be the rump of these companies, firmly entrenched within the OECD 'laager', which will continue in business, diversifying into a range of newer industries, such as petrochemicals or offshore mineral extraction. Within the Western industrialised world, they will tend to own their business operations outright and will thus remain recognisable as the type of multinational companies to which we have become accustomed – though there will be some deviations from this model. For one thing, there will be a growing foreign stake in such companies. During 1976, Iran showed serious interest in buying an equity stake in both BP and Occidental[5] and one or two of the smaller independents may well be acquired by national oil companies of the size of Petromin or NIOC. It is less likely that this will happen to one of the majors, given the unwillingness of Western governments to let companies of such central strategic importance fall into the hands of foreign states, but circumstances might arise whereby producer states might be allowed to take a majority stake in certain subsidiary companies, or a carefully defined minority stake in one or two of the majors.

Within the OECD area, the growing concern with security of supplies may lead national governments to encourage various subsidiaries of the majors to seek a greater degree of autonomy from each other and from international headquarters. This will be nothing new for a company like Shell, which has long been used to the fact that its US subsidiary has been, for antitrust purposes, very much an independent entity. However, all companies will have to grow used to a situation in which, say, their German subsidiaries may sign an agreement with the German government to develop sources of oil earmarked specifically for the German market, whatever the overall circumstances of the parent company. On the other hand, even though nationalist pressures within the OECD area may force company subsidiaries to think more parochially, it is doubtful if such pressures will become overbearing. Doubtless international headquarters will give national subsidiaries a greater say in decisions affecting themselves and more freedom to compete with other subsidiaries for third markets, but this is a trend which is affecting virtually all multinationals – not just the oil companies.

The 'core' companies will also enter into a new variety of relationships with governments, national champions and smaller private companies, whose pattern will reflect both old loyalties and current strengths in various technological and geographical areas. They will probably develop a variety of relationships, starting with semi-permanent joint ventures, through fixed-term management contracts to once-and-for-all commercial exchanges of goods and services. Even in these latter shorter-term transactions, there will often be continuity with large-scale purchasers of goods and services, generally the national oil companies, tending to favour a limited number of suppliers, generally the traditional companies, out of convenience and inertia. The core companies will still exert a fair amount of influence over newer entrants into the industry although these will probably never be clear-cut patterns of interdependence like those found in other industries such as the international airline industry, where world airlines are dependent on aircraft from specific manufacturers like Boeing, McDonnell Douglas or Lockheed. Oil technologies are not likely to be monopolised by a handful of companies although the latter will remain strong enough in a variety of technologies and markets to entice a significant number of smaller companies into a loosely structured orbit around them. These groupings may well be similar to the informally linked companies found in the USA and Germany grouped round certain dominant families or banks, or to the constellation of Japanese companies found round each of Japan's giant trading companies. In the former case, the family bonds are relatively weak, resting on a financial control which is exercised primarily in times of crisis. The oil company connections will be formed of much stronger links of management intervention and long-run commercial transactions. In the Japanese case, the bonds are historical and accepted quite happily by all the family members. In the oil industry, many of the satellites will be newly formed national oil companies willing to associate only if the economic incentive is particularly convincing. They will not deal exclusively with the majors, but, particularly in the face of complex industrial problems, are more likely to collaborate with them than not.

We are still very close to the heady days of the 1973 oil embargo, and the final participation deals involving key companies like Aramco had not been signed when this book went to press. However, whatever the final relationship between Aramco and the Saudi authorities, it seems likely that there will be a fairly lasting interaction between the Saudi oil industry and at least some of the old Aramco partners, though there will obviously be a number of new entrants clamouring to challenge their pre-eminent position. The fact that a company like NIOC is involved in offshore exploration in the North Sea in collaboration with BP suggests that, just as Shell and Exxon have had a close relationship in similar exploration over the past years, national oil companies will come to form semi-permanent working relations with various private companies.

The conclusion that the traditional oil companies will continue to survive and prosper will not be supported by those ecologists who believe that the world will run out of oil in the early years of the twenty-first century. It will not please those who believe in the attainability of the New International Economic Order, who see no reason why Third World governments controlling the world's oil supplies should not sweep the majors into the dustbin of history, or left-wing radicals who hope or believe that such companies will everywhere be dismembered or nationalised on ideological grounds.

The case against the depletion arguments of the Club of Rome is that the survival of the oil companies does not depend in any way on the accuracy of the doomsters' predictions as they are diversifying into other fields. Should the pessimists prove to be right, the search for alternative sources of energy would become more hectic – a situation which would favour the leading oil companies, whose strategies depend on identifying fast-moving areas of technology to which they can apply their financial and managerial strengths. Paradoxically, the position of the companies would be worse if the doomsters prove to be wrong and supplies of oil last well into the next century, for, in those circumstances, the technology of oil production would favour national oil companies which can rely on political favouritism.

The case against the hopes of the Third World enthusiast is that the newer national oil companies will not have the depth of managerial experience to be fully competitive with the majors in free competition, at least for a while. In fact they may actually choose not to compete but concentrate on the industrialisation of their home economies. The majors still control the most important ultimate markets and industrialised governments will have security objections to increasing their dependence on OPEC-based companies. There are both commercial and political doubts about Third World optimism that the traditional companies will become of little importance.

Finally, in answer to Western radicals, there is little sign that the majors will disappear as private entities in the short to medium term as a result of nationalisation or divestiture within the OECD area. Certainly, governments have become increasingly sensitive to oil issues, but, with the exception of the Communist Party in France (which in 1977 was calling for the nationalisation of CFP), there appear to be no parties in the OECD area supporting nationalisation which have a realistic chance of coming to power. The relevant governments seem to have sought international solutions to security issues through bodies like the International Energy Agency and, when they have felt the need to take a more direct role in their domestic industries, have created state companies as parallel institutions. The complexity of the jurisdictional problems which would be thrown up by nationalisations or divestitures would certainly discourage such actions within the OECD, especially in an era in which international uncertainty makes the traditional oil companies useful buffer organisations for the industrialised world when facing the OPEC nations.

Perhaps the most interesting long-term institutional development affecting the private oil companies would be the emergence of some form of international supervision over their activities. This will be a gradual process working at several levels. Like all multinationals, oil companies will become increasingly subject to tougher national regulations, standardised through bodies like the OECD and EEC, for the industrialised world, and UNCTAD and the UN Centre on

Transnational Corporations, for the Third World. These organisations will have to fight very hard to exert supranational authority in an area of such great national concern but all the indications are that they will become important centres of influence over national policy makers in the regulation of such companies.

Unlike their manufacturing counterparts, who will continue to do business within a framework of national jurisdictions, the oil companies (along with a limited number of others) will also find themselves working directly under whatever international agency is set up to supervise the exploitation of the deep seabed. It is too soon to say what form this body will take but as the first major operational international agency created since the upsurge of Third World influence within world councils during the 1960s, it is more likely to reflect the attitudes of the average UN member than those of traditional parent governments. The international deep seabed authority will be an obvious means to extend the international community's authority over such companies, since it will control access to a commercially valuable domain which will feature heavily in the latter's longer-term planning. It may well be that we never reach the stage foreseen by some visionaries in which the multinationals register directly with some supranational authority, but it seems eminently possible that the deep sea authority will insist that companies operating in its area should do so through specially constituted and registered subsidiaries, perhaps with some direct form of representation by the international community in their management. As it is now accepted that an oil-producing state like Iran may try to buy a stake in a sizeable company like Occidental, it is only one stage further for parts of the UN family, which are now increasingly coming to represent the interests of the poorer world, to strive to win themselves a similar share in management of the world economy. If this does indeed happen, it will be the third major revolution to have affected the oil industry this century (the first having been the American incursion into European imperial preserves, the second, the host governments' challenge to the industrialised world's hegemony). However, any future change will be much more gradual, since supranational authorities will not be able to call on the diplomatic resources of a superpower, nor on the unquestioned jurisdictional rights of even the smallest sovereign state. Instead, such an authority will have to win general acceptance and, to do this, will have to live with initially conflicting demands from the industrialised and developing worlds. Gradually, though, by rejecting companies which insist on unreasonable terms for key concessions, and by building up its own institutions capable of exploiting major mineral deposits in their own right, it will reach a situation where the companies need it more than it needs any specific company.

At that point, this supranational authority will have come of age.

The future of the former private oil companies will be one of steady evolution. Certainly the 1970s have produced a short, sharp redefinition of their role, but from now on they will be moving to consolidate their hold on the commercial openings left by governmental planners. These companies will survive, providing they can transfer the commercial lessons of one part of the world to other parts of the globe. The general atmosphere towards their activities will be less friendly and more challenging. They will have to protect themselves by widening the scope of their transnational alliances. They will have to count far more on their commercial acumen than on the political clout of their parent governments. Some will undoubtedly fail to adjust fast enough and will disappear as separate entities. However, the bulk of the progeny of pioneers such as John D. Rockefeller, Sir Henri Deterding, and William Knox D'Arcy will survive as recognisable entities well into the twenty-first century.

NOTES

1. Odell argues that it is unfair to use the Pertamina case as an example of the workings of national oil companies. He would prefer to point to the relatively smooth running of Pemex, Petrobras and ENAP.
2. Odell informs me that Petrobras has the world's first successful offshore subsea completion system and has a patented oil-from-shales process.
3. It has been suggested that a possible Iranian 500,000 barrels-a-day export refinery could only deliver oil products competitively to Japan if the crude oil were fed into it at $4.70 a barrel under the 1976 OPEC price (*Oil and Gas Journal,* 28 June 1976).
4. Stopford and Wells (1972, p. 64) and Franko (1971, p. 90) both show that many manufacturing companies have moved on to still further types of organisational structure.
5. Neither of these deals came off, but rumours of a link between Iran's NIOC and Italy's ENI remain. Even if this deal does not come off, some others will eventually.

Bibliography

Aamo, Bjorn Skogstad, 'Norwegian oil policy: basic objectives', in *The Political Implications of North Sea Oil and Gas*, ed. Martin Saeter and Ian Smart (Guildford: IPC Science and Technology Press, 1975), pp. 81–92.

Adelman, M. A., *The World Petroleum Market* (Baltimore: Johns Hopkins University Press, 1972).

Adelman, M. A., 'Is the oil shortage real? Oil companies as OPEC tax-collectors', *Foreign Policy* (Winter 1972–3), pp. 69–107.

Akins, James E., 'The oil crisis: this time the wolf is here', *Foreign Affairs*, vol. 51, no. 3 (1973), pp. 462–90.

Allen, Harry Cranbrook, *Great Britain and the United States: A History of Anglo-American Relations (1783–1952)* (London: Odhams, 1954).

Alperovitz, Gar, *Atomic Diplomacy: Hiroshima and Potsdam, the Use of the Atomic Bomb and the American Confrontation with Soviet Power* (New York: Simon & Schuster, 1965).

Anderson, Irvine H., *The Standard-Vacuum Oil Company and United States East Asian Policy 1933–1941* (Princeton, NJ: Princeton University Press, 1975).

Aruri, Naseer H. and Hevener, Natalie K., 'France and the Middle East', in *The Middle East in World Politics*, ed. Tareq Y. Ismael (Syracuse, NY: Syracuse University Press, 1974), pp. 59–93.

Badeau, John S., *The American Approach to the Arab World* (New York: Harper & Row, 1968).

Ball, George W. (ed.), *Global Companies: The Political Economy of World Business* (Englewood Cliffs, NJ: Prentice-Hall, 1975).

Bauer, Raymond A., Pool, Ithiel de Sola and Dexter, Lewis A., *American Business and Public Policy: The Politics of Foreign Trade* (New York: Atherton, 1963).

Blair, John M., *The Control of Oil* (London: Macmillan, 1976).

Boddewyn, J. and Kapoor, Ashok, 'The external relations of American multinational enterprises', *International Studies Quarterly*, vol. 16, no. 4 (1972), pp. 433–53.

Bonsal, Philip W., *Cuba, Castro and the United States* (Pittsburgh: University of Pittsburgh Press, 1971).

BP 1956, 1973, 1974, 1976, *BP Statistical Review of the World Oil Industry* (London: BP).

Cable, James, *Gunboat Diplomacy: Political Applications of Limited Naval Force* (London: Chatto & Windus, 1971).

Caldwell, Malcolm, *Oil and Imperialism in East Asia* (Nottingham: Bertrand Russell Peace Foundation, 1971).

Callcott, Wilfrid Hardy, *The Western Hemisphere: Its Influence on United States Policies to the end of World War II* (Austin: University of Texas Press, 1968).

Calvert, Peter, *The Mexican Revolution 1910–1914: The Diplomacy of Anglo-American Conflict* (Cambridge: Cambridge University Press, 1968).
Caroe, Olaf, *Wells of Power: The Challenge to Islam: A Study in Contrast* (London: Macmillan, 1951).
Chandler, Geoffrey, 'The changing state of the oil industry', *Petroleum Review* (June 1974), pp. 375–81.
Chase Manhattan, *Capital Investments of the World Petroleum Industry* (New York: Chase Manhattan, 1976).
Chenery, Hollis *et al., Redistribution with Growth* (London: Oxford University Press, 1974).
Cheney, Michael Sheldon, *Big Oilman from Arabia* (London: Heinemann, 1958).
Chevalier, Jean-Marie, *The New Oil Stakes* (London: Allen Lane, 1975).
Childers, Erskine B., *The Road to Suez: A Study of Western-Arab Relations* (London: MacGibbon & Kee, 1962).
Chisholm, A. H. T., *The First Kuwait Oil Agreement: A Record of Negotiations, 1911–1934* (London: Frank Cass, 1975).
Church, Frank, *The International Petroleum Cartel, the Iranian Consortium and US National Security,* prepared for the use of the Sub-Committee on Multinational Corporations of the Committee on Foreign Relations, US Senate (Washington: US Government Printing Office, 1974).
Church, Frank, *Multinational Petroleum Companies and Foreign Policy,* hearings before the Sub-Committee on Multinational Corporations of the Committee on Foreign Relations, US Senate, vols 4, 5, 6 and 7 published 1974, vol 8 published 1975 (Washington: US Government Printing Office).
Church, Frank, *Multinational Oil Corporations and US Foreign Policy,* report together with individual views to the Committee on Foreign Relations, US Senate, by the Sub-Committee on Multinational Corporations (Washington: US Government Printing Office, 1975).
Church, Frank, *Political Contributions to Foreign Governments,* hearings before the Sub-Committee on Multinational Corporations of the Committee on Foreign Relations, US Senate, pt 12 (Washington: US Government Printing Office, 1976).
Copeland, Miles, *The Game of Nations* (London: Weidenfeld & Nicolson, 1969).
Cronje, Suzanne, *The World and Nigeria: The Diplomatic History of the Biafran War 1967–1970* (London: Sidgwick & Jackson, 1972).
Darmstadter, Joel and Landsberg, Hans H., 'The economic background', *Daedalus,* vol. 104, no. 4 (1975), pp. 15–37.
Debanné, J. G., 'Oil and Canadian policy', in *The Energy Question: An International Failure of Policy; Vol. 1: The World; Vol. 2: North America,* ed. Edward W. Erickson and Leonard Wavermann (Toronto: University of Toronto Press, 1974), pp. 125–48.
Denny, Ludwell, *We Fight for Oil* (New York: Alfred A. Knopf, 1929).

Denny, Ludwell, *America Conquers Britain: A Record of Economic War* (London: Alfred A. Knopf, 1930).

De St Jorre, John, *The Nigerian Civil War* (London: Hodder & Stoughton, 1972).

Domhoff, G. William, 'Who made American foreign policy 1945–1963?', in *Corporations and the Cold War*, ed. David Horowitz (New York: Monthly Review Press, 1969), pp. 25–69.

Eden, Anthony, *Full Circle* (London: Cassell, 1960).

Elwell-Sutton, L. P., *Persian Oil: A Study in Power Politics* (London: Lawrence & Wishart, 1955).

Engler, Robert, *The Politics of Oil* (New York: Macmillan, 1961).

Erickson, Edward W. and Waverman, Leonard (eds), *The Energy Question: An International Failure of Policy; Vol. 1: The World; Vol. 2: North America* (Toronto: University of Toronto Press, 1974).

Exxon, *Middle East Oil,* Exxon background series (New York: Exxon, 1976).

FEA (Federal Energy Administration), *The Relationship of Oil Companies and Foreign Governments* (Washington: FEA, 1975a).

FEA, *US Oil Companies and the Arab Oil Embargo: The International Allocations of Constricted Supplies,* prepared for the Sub-Committee on Multinational Corporations of the Committee on Foreign Relations, US Senate (Washington: US Government Printing Office, 1975b).

Fesharaki, Fereidun, *Development of the Iranian Oil Industry: International and Domestic Aspects* (New York: Praeger, 1976).

Fifer, J. Valerie, *Bolivia: Land, Location and Politics since 1825* (Cambridge: Cambridge University Press, 1972).

First, Ruth, *Libya: The Elusive Revolution* (Harmondsworth: Penguin, 1974).

Fleming, D. F., *The Cold War and its Origins 1917–1960* (London: Allen & Unwin, 1961).

Forbes, 'Don't blame the oil companies: blame the State Department', *Forbes* (15 April 1976), pp. 69–85.

Frankel, P. H., *Mattei: Oil and Power Politics* (London: Faber & Faber, 1966).

Franko, Lawrence G., *Joint Venture Survival in Multinational Corporations* (New York: Praeger, 1971).

Galbraith, John Kenneth, 'Into the sunset', *New York Review of Books* (18 March 1976), pp. 10–14.

Gardner, Lloyd C., *Architects of Illusion: Men and Ideas in American Foreign Policy 1941–1949* (Chicago: Quadrangle, 1970).

Ghadar, Fariborz, 'A study in the evolution of strategy in raw material exporting nations', thesis submitted in partial fulfilment of requirements for degree of Doctor of Business Administration, Graduate School of Business, Harvard University (mimeo., 1975).

Gilpin, Robert, *France in the Age of the Scientific State* (Princeton, NJ: Princeton University Press, 1968).

Gilpin, Robert, *US Power and the Multinational Corporation: The Political Economy of Foreign Direct Investment* (London: Macmillan, 1976).

Goodsell, Charles T., *American Corporations and Peruvian Politics* (Cambridge, Mass.: Harvard University Press, 1974).
Hartley, 1956: see OEEC, 1956.
Hartshorn, J. E., *Oil Companies and Governments: An Account of the International Oil Industry in its Political Environment,* 2nd edn (London: Faber & Faber, 1967).
Henriques, Robert, *Marcus Samuel* (London: Barrie & Rockliff, 1960).
Hexner, Ervin, *International Cartels* (London: Pitman & Sons, 1946).
Hirst, D., *Oil and Public Opinion in the Middle East* (London: Faber & Faber, 1966).
Hobsbawm, E. J., 'Dictatorship with charm', *New York Review of Books,* vol. 22, no. 15 (1975), pp. 22–4.
Hodges, Michael, *Multinational Corporations and National Government* (Farnborough: Saxon House, 1974).
Hoffmann, Stanley, 'International organization and the international system', *International Organization,* vol. 24, no. 3 (1970).
Horowitz, David, *The Free World Colossus: A Critique of American Foreign Policy in the Cold War* (New York: Hill & Wang, 1965).
Horowitz, David (ed.), *Corporations and the Cold War* (New York: Monthly Review Press, 1969).
Howell, Leon and Morrow, Michael, *Asia, Oil Politics and the Energy Crisis: The Haves and the Have-Nots* (New York: IDOC/North America, 1974).
Ismael, Tareq Y., *The Middle East in World Politics* (Syracuse, NY: Syracuse University Press, 1974).
Issawi, Charles and Yeganeh, Mohammed, *The Economics of Middle Eastern Oil* (London: Faber & Faber, 1963).
Jacoby, Neil H., *Multinational Oil* (New York: Macmillan, 1974).
Jay, Antony, *Management and Machiavelli: An Inquiry into the Politics of Corporate Life* (London: Hodder & Stoughton, 1967).
Kent, Marian, *Oil Empire: British Policy and Mesopotamian Oil 1900–1920* (London: Macmillan, 1976).
Keohane, Robert O. and Nye, Joseph S., Jnr, *Power and Interdependence* (Boston: Little, Brown, 1977).
Keohane, Robert O. and Ooms, Van Doorn, 'The multinational firm and international regulations', *International Organization,* vol. 29, no. 1 (1975), pp. 169–200.
Kindleberger, Charles P., *American Business Abroad: Six Lectures on Direct Investment* (New Haven, Conn.: Yale University Press, 1969).
Kolko, Gabriel, *The Politics of War: The World and United States Foreign Policy 1943–5* (New York: Vintage Books, 1968).
Kolko, Gabriel, *The Roots of American Foreign Policy: An Analysis of Power and Purpose* (Boston: Beacon Press, 1969).
Landis, Lincoln, *Politics and Oil: Moscow in the Middle East* (New York: Donellen, 1973).
Lantzke, Ulf, 'The OECD and its International Energy Agency', *Daedalus,* vol. 104, no. 4 (1975), pp 217–27.
Lenczowski, George, *The Middle East in World Affairs,* 3rd edn (Ithaca: Cornell University Press, 1962).

Levitt, Kari, *Silent Surrender: The American Economic Empire in Canada* (New York: Liveright, 1970).

Levy, Walter J., 'Oil power', *Foreign Affairs,* vol. 49, no. 4 (1971), pp. 652–68.

Levy, Walter J., 'An Atlantic-Japanese energy policy', paper delivered to the Europe-American conference: 'New Roles and Relationships in the Next Decade': Amsterdam, March 1973 (*Petroleum Press Service,* April 1973, pp. 127–9).

Lieuwen, Edwin, *Petroleum in Venezuela: A History* (Berkeley: University of California Press, 1954).

Lieuwen, Edwin, *Venezuela,* 2nd edn (London: Oxford University Press, 1965).

Lindsay, J. W., 'The war over the Chaco', *International Affairs,* vol. 14 (1935), pp. 231–40.

Litvak, Isiah and Maule, Christopher, 'Nationalisation in the Caribbean bauxite industry', *International Affairs,* vol. 51, no. 1 (1975), pp. 43–59.

Longhurst, Henry, *Adventure in Oil: The Study of British Petroleum* (London: Sidgwick & Jackson, 1959).

Longrigg, Stephen Hemsley, *Oil in the Middle East: Its Discovery and Development,* 3rd edn (London: Oxford University Press, 1968).

Maddox, Robert James, *The New Left and the Origins of the Cold War* (Princeton, NJ: Princeton University Press, 1973).

McFadzean, Frank S., 'The role of the oil companies', an address at Harvard University, 20 March 1973.

McKie, James W., 'The United States', *Daedalus,* vol. 104, no. 4 (1975), pp. 73–90.

Madelin, Henri, *Oil and Politics* (Farnborough: Saxon House, 1975).

Meadows, Donella H. *et al.*, *The Limits to Growth* (London: Earth Island, 1972).

Mendershausen, Horst, *Coping with the Oil Crisis: French and German Experiences* (Baltimore: Johns Hopkins University Press, 1976).

Mendle, Wolf, *Issues in Japan's China Policy* (London: Macmillan, 1978).

Monroe, Elizabeth, *Britain's Moment in the Middle East* (London: Chatto & Windus, 1963).

Moran, Theodore H., *Multinational Corporations and the Politics of Dependence* (Princeton, NJ: Princeton University Press, 1974).

Mosley, Leonard, *Power Play: The Tumultuous World of Middle East Oil 1890–1973* (London: Weidenfeld & Nicolson, 1973).

Nation, 'Blood and oil: Aramco's secret report on Palestine', *Nation* (June 26, 1948), pp. 705*ff*.

Nearing, Scott and Freeman, Joseph, *Dollar Diplomacy* (New York: Monthly Review Press, 1969).

Nutting, Anthony, *No End of a Lesson: The Story of Suez* (London: Constable, 1967).

Nye, Joseph S., Jnr, 'Multinational corporations and world order', in *Global Companies: The Political Economy of World Business,* ed. George W. Ball (Englewood Cliffs, NJ: Prentice-Hall, 1975), pp. 122–47.

Nye, Joseph S., Jnr and Keohane, Robert O., *Transnational Relations and World Politics* (Cambridge, Mass.: Harvard University Press, 1973).

O'Connor, Richard, *The Oil Barons: Men of Greed and Grandeur* (London: Hart-Davis/MacGibbon, 1972).

Odell, Peter R., 'The oil industry in Latin America', in *The Large International Firm in Developing Countries: The International Petroleum Industry*, ed. Edith Penrose (London: Allen & Unwin, 1968), pp. 274–300.

Odell, Peter R., *Oil and World Power: A Geographical Interpretation,* 1st edn (Harmondsworth: Penguin, 1970); 2nd edn (Harmondsworth: Penguin, 1972).

OEEC (Organisation of Europeon Economic Co-operation), *Europe's Growing Needs of Energy: How can they be met?* (Hartley Report) (Paris: OEEC, 1956).

OEEC, *Europe's Need for Oil: Implications and Lessons on the Suez Crisis* (Paris: OEEC, 1958).

OEEC, *Towards a New Energy Pattern in Europe* (Robinson Report) (Paris: OEEC, 1960).

OECD (Organisation of Economic Co-operation and Development), *Energy Prospects to 1985* (Paris: OECD, 1974).

Oppenheim, V. H., 'The past: we pushed them', *Foreign Policy*, no. 25 (Winter 1976–7), pp. 24–57.

Payton-Smith, D. J., *History of the Second World War: Oil, a Study of War-time Policy and Administration* (London: HMSO, 1971).

Pearson, Scott R., *Petroleum and the Nigerian Economy* (Stanford: Stanford University Press, 1970).

Pearton, Maurice, *Oil and the Romanian State* (Oxford: Clarendon Press, 1971).

Pendle, George, *Paraguay: A Riverside Nation* (London: Oxford University Press, 1968).

Penrose, Edith (ed.), *The Large International Firm in Developing Countries: The International Petroleum Industry* (London: Allen & Unwin, 1968).

Penrose, Edith, *The Growth of Firms, Middle East Oil and Other Essays* (London: Frank Cass, 1971).

Penrose, Edith, 'The development of crisis', *Daedalus*, vol. 104, no. 4 (1975), pp. 39–57.

Pinelo, A. J., *The Multinational Corporation as a Force in Latin American Politics: A Case Study of the International Petroleum Company in Peru* (New York: Praeger, 1973).

Prodi, Romano and Clô, Alberto, 'Europe', *Daedalus*, vol. 104, no. 4 (1975), pp. 91–112.

Rand, Christopher T., *Making Democracy Safe for Oil: Oilmen and the Islamic East* (Boston: Little, Brown, 1975).

Ridgeway, James, *The Last Play: The Struggle to Monopolise the World's Energy Resources* (New York: Mentor, 1973).

Rifäi, Taki, 'La crise petrolière internationale (1970–1971): essai d'interprétation', *Revue Française de Science Politique,* vol. 22, no. 6 (1972), pp. 1205–36.

Robinson Report 1960: see OEEC, 1960.

Rondot, Jean, *La Compagnie Française des Pétroles* (Paris: Plon, 1962).

Rustow, Dankwart and Mugno, John F., *OPEC: Success and Prospects* (New York: New York University Press, 1976).

Sale, Kirkpatrick, *Power Shift: The Rise of the Southern Rim and its Challenge to the Eastern Establishment* (New York: Random House, 1975.

Sampson, Anthony, *The Seven Sisters: The Great Oil Companies and the World They Made* (London: Hodder & Stoughton, 1975).

Sampson, Anthony, 'Statement to the US Senate Sub-Committee on Antitrust and Monopoly' (mimeo, 3 February 1976).

Scott, George, *The Rise and Fall of the League of Nations* (London: Hutchinson, 1973).

Shell, *Oil and Gas in the Enlarged European Community* (London: Shell, 1972).

Shell, *Oil and Gas in 1974* (London: Shell, 1975).

Shwadran, Benjamin, *The Middle East, Oil and the Great Powers,* 3rd edn (New York: John Wiley, 1973).

Smith, David N. and Wells, Louis T., Jnr, 'Mineral agreements in developing countries: structures and substance', *American Journal on International Law,* vol. 69, no. 3 (1975), pp. 560–90.

Stobaugh, Robert B., 'The oil companies in the crisis', *Daedalus,* vol. 104, no. 4 (1975), pp. 179–202.

Stobaugh, Robert B., 'Petrochemicals and the economics of integration', lecture at St Antony's College, Oxford, 10 February 1976.

Stopford, John M. and Wells, Louis T. Jnr, *Managing the Multinational Enterprise: Organisation of the Firm and Ownership of the Subsidiaries* (London· Longman, 1972).

Straszheim, Mahlon R., *The International Airline Industry* (Washington DC: Brookings Institute, 1969).

Sutton, Anthony C., *Western Technology and Soviet Economic Development 1917 to 1930* (Stanford: Stanford University Press, 1968).

Tanzer, Michael, *The Political Economy of International Oil and the Underdeveloped Countries* (London: Temple-Smith, 1969).

Tarbell, Ida, *The History of the Standard Oil Company* (New York: Macmillan, 1904).

Taylor, Wayne Chatfield and Lindeman, John, *Creole Petroleum Corporation in Venezuela* (Washington DC: National Planning Association, 1955).

Thomas, Hugh, *Cuba or the Pursuit of Freedom* (London: Eyre & Spottiswood, 1971).

Toynbee, Arnold J., *Survey of International Affairs 1928* (London: Oxford University Press, 1929).

Trumpener, Ulrich, *Germany and the Ottoman Empire 1914–1918* (Princeton, NJ: Princeton University Press, 1968).

Tsurumi, Yoshi, 'Japan', *Daedalus*, vol. 104, no. 4 (1975), pp. 113–27.

Tugendhat, Christopher and Hamilton, Adrian, *Oil the Biggest Business*, rev. edn (London: Eyre Methuen, 1975).

Tugwell, Franklin, *The Politics of Oil in Venezuela* (Stanford: Stanford University Press, 1975).
Turner, Louis, *Multinational Companies and the Third World* (New York: Hill & Wang, 1973).
Turner, Louis, 'The European Community; features of disintegration: politics of the energy crisis', *International Affairs,* vol. 50, no. 3 (1974a), pp. 404–15.
Turner, Louis, 'The Washington Energy Conference', *The World Today,* vol. 30, no. 3 (1974b), pp. 89–92.
Turner, Louis, 'State and commercial interests in North Sea Oil and Gas: conflict and correspondence', in *The Political Implications of North Sea Oil and Gas*, ed. Martin Saeter and Ian Smart (Guildford: IPC Science & Technology Press, 1975), pp. 93–110.
Vallenilla, Luis, *Oil: The Making of a New Economic Order: Venezuelan Oil and OPEC* (New York: McGraw-Hill, 1975).
Vernon, Raymond, *Sovereignty at Bay: The Multinational Spread of US Enterprises* (London: Longman, 1971).
Vernon, Raymond (ed.), *Big Business and the State: Changing Relations in Western Europe* (London: Macmillan, 1974).
Walton, Ann-Margaret, 'Atlantic bargaining over energy', *International Affairs,* vol. 52, no. 2 (1976), pp. 180–96.
Waverman, Leonard, 'The reluctant bride: Canadian and American energy relations', in *The Energy Question: An international failure of policy,* ed. Edward W. Erickson and Leonard Waverman (Toronto: University of Toronto Press, 1974), pp. 217–38.
Wilkins, Mira, *The Maturing of Multinational Enterprise: American Business Abroad from 1914–1970* (Cambridge, Mass.: Harvard University Press, 1974).
Wilkins, Mira, 'The oil companies in perspective', *Daedalus,* vol. 104, no. 4 (1975), pp. 159–78.
Williams, William Appleman, *The Tragedy of American Diplomacy* (New York: Dell Publishing, 1962).
Williamson, J. W., *In a Persian Oil Field: A Study in Scientific and Industrial Development* (London: Ernest Benn, 1927).
Wood, Bryce, *The United States and Latin American Wars 1932–1942* (New York: Colombia University Press, 1966).

Index

(Compiled by Mary Hargreaves)